POLISH CONTRIBUTIONS TO THE SCIENCE OF SCIENCE

THEORY AND DECISION LIBRARY

AN INTERNATIONAL SERIES
IN THE PHILOSOPHY AND METHODOLOGY OF THE
SOCIAL AND BEHAVIORAL SCIENCES

VOLUME 29

POLISH CONTRIBUTIONS TO THE SCIENCE OF SCIENCE

edited by

BOHDAN WALENTYNOWICZ
Polish Academy of Sciences

D. REIDEL PUBLISHING COMPANY
DORDRECHT : HOLLAND / BOSTON : U.S.A.
LONDON : ENGLAND

PWN – POLISH SCIENTIFIC PUBLISHERS
WARSZAWA

Library of Congress Cataloging in Publication Data

CIP

Main entry under title:
Polish contributions to the science of science.
(Theory and decision library; v. 29)
Indudes index.
1. Science—Addresses, essays, lectures.
I. Walentynowicz, Bohdan. II. Series.
Q171.P773 500 81-23415
ISBN 90-277-1233-6 AACR2

Distributors for Albania, Bulgaria, Chinese People's Republic, Cuba, Czechoslovakia, German Democratic Republic, Hungary, Korean People's Democratic Republic, Mongolia, Poland, Romania, the U.S.S.R., Vietnam, and Yugoslavia

ARS POLONA

Krakowskie Przedmieście 7, 00-068 Warszawa 1, Poland

Distributors for the U.S.A. and Canada
by KLUWER BOSTON INC.,
190 Old Derby Street, Hingham, MA 02043, U.S.A.

Distributors for all other countries
KLUWER ACADEMIC PUBLISHERS GROUP,
P.O. Box 322, 3300 AH Dordrecht, Holland
D. Reidel Publishing Company is a member of the Kluwer Group

Printed in Poland

TABLE OF CONTENTS

PREFACE

The present volume helps to understand how the collection of miscellaneous disciplines, called the science of science, came into being.

One tradition goes back to the Greek thinkers and includes practically every great name in the history of philosophy, since almost all philosophers took a stand on epistemological problems. To be sure, theory of knowledge is not to be identified with theory of science. Science is a relatively new phenomenon, dating from modern times. It is a source of knowledge, the main source (in our type of culture), but certainly not the only one.

Also, epistemology deals with questions rather far removed from the practice of science. The nature of truth; the sources of knowledge; the limits of knowledge. Many people claim to understand science without being able to solve these perennial problems. What we call today the science of science is not concerned with, say, the Kantian question how synthetic *a priori* truths are possible.

However, the influence of the philosophical heritage is still felt in present day metascientific thought. Philosophy of science does exist, although in a form very different from classical epistemology. It has developed (under that name) into general methodology of science, which roughly means: analysis of scientific theories and analysis of their validation. In both cases, mathematical and logical tools play a great role. Scientific theories are analyzed by means of modern semantics (theory of models). Their validation is reconstructed in terms of probability and decision theory. Ideally, the methodology of justification of scientific theories follows the program outlined by Kazimierz Ajdukiewicz (1890–1963): to abstract from scientific practice the patterns actually followed and to show their efficiency with respect to a (hypothetical) goal. In this way, scientists following these patterns of behaviour are shown to be rational agents.

The philosophical origin of metascience is easily observed in the pioneering writings of Znaniecki, Kotarbiński, Maria Ossowska and Stanisław Ossowski, dating mostly from the inter-war period.

So much has been written on the subject in the meantime that the very idea of treating science as a social and cultural phenomenon appears commonplace. Fifty years ago it was considered quite revolutionary and called for some explanation. Hence the apologetic tone of certain writings from that period. M. and S. Ossowski think it necessary to defend the anthropological approach to science by recalling that science lends to contemporary culture its unity and universality, while Znaniecki points out that we could arrive at a better understanding of science if we knew more about its relation to the social context.

A new impetus was given to such investigations when it was realized that science is a social enterprise too big and too important to be allowed to develop spontaneously. A famous book by J. D. Bernal (*The Social Function of Science*, 1939) marks the beginning of large-scale research on science. It began with practical problems, such as the organization and planning of science, allocation of resources, utilization of results, etc., but very quickly there appeared sociological and psychological problems which were interesting in their own right. How and why do scientific schools come into being? What is the role of authority in science? Does science create its own ethical problems? What are the conditioning factors of discovery? Can scientific talent be diagnosed? Thus the science of science, postulated in the twenties, has become a reality in the post-war period.

At the same time, it began to suffer disintegration. Obviously, there is little connection between, say, the economics of science and the structure of scientific theory. An expert on the legal aspects of science can hardly find a common language with a decision theorist who develops a model of scientific research as a game against nature.

In the paper by M. and S. Ossowski this possibility was already taken into account. They asked, "can there exist a specialist of the science of science, a specialist who would have to be experienced in general methodology, to be familiar with the achievements of psychology, of the theory and history of culture, etc.?". Of course not. They argue (correctly) that the situation is similar in other disciplines, e.g. in linguistics, where apparantly it does no harm. However, the paper was written before the time when specialization was pushed so far that it did become harmful.

The need for integration is nowadays generally acknowledged. Here is one example. For at least twenty years we have noted a growing dissatisfaction with the results of philosophy of science. The main objection is that they are far removed from the practice of science. Philosophy of science works with simple models which are openly unrealistic. Their analogy with actual scientific theories is a matter of faith rather than argument. Attempts to model parts of real science have, so far, succeeded on an extremely small scale and it is doubtful whether these models can be extended.

Dissatisfaction with the traditional philosophy of science led to a change in the point of view. Whereas philosophy of science concentrated on finished scientific theories and the procedures validating them, according to the new approach it is *change* that constitutes the focal point of interest. Such an attitude was already present in the writings of Karl Popper (*Logik der Forschung*, 1934) and became widely adopted after the publication of *The Structure of Scientific Revolutions* by T. S. Kuhn. Instead of artificial models, it was real science that was the object of investigation. History and sociology of science replaced logical analysis.

This trend has brought valuable findings concerning the development of science. Its shortcomings, however, have by now become apparent. It is obvious that to present scientific research as a kind of game which scientists play among themselves is to lose from view the point of science. Whether we call it, in the old-fashion way, the search for truth or, more modestly, the search for adequacy, science develops under the control of an external factor — reality, which somehow must be taken into account.

It thus seems that what the science of science urgently needs is some kind of integration of the two trends, which at present develop in parallel, with hardly any connection between them.

There are, of course, considerable difficulties, due mainly to the difference in methods. But the job will have to be done.

Another problem of integration (or is it just a variant of the above one?) is connected with the well known distinction, due to Hans Reichenbach, between methods of discovery and methods of justification. The disctinction seemed to point to an unbridgeable gap, as long as methods of discovery were so vague that they hardly deserved to be called methods. But the recent results in heuristics, particularly computer-modelling of the search for solution, suggest that we can at least look for a connection between the two kinds of methods. Discovery does presuppose justification, otherwise why should it be called discovery at all? How does this fact influence the methods?

If we are to counteract the undesirable consequences of specialization, it is perhaps a good policy to go back to the sources, i.e. to the time when the unity due to the common subject of investigation had more weight than the differences in approach. The history of the science of science may provide a clue as to how integration could be achieved. Or, rather, regained.

Klemens Szaniawski

EDITOR'S NOTE

Every book has its own genesis. What is the genesis of this book?

Throughout the last ten or more years both European as well as American authors in their reflections about science have made references to a paper by the Polish sociologists Stanisław Ossowski and Maria Ossowska, entitled 'Science of Science', the English version of which was published in the periodical *Nauka Polska*, No. X, Vol. 3, 1935, and have emphasized its pioneer character. This was the first time that the term 'science of science' was used, and problems with which this discipline was concerned were delineated, since they differ both from those dealt with by the 'philosophy of science', and those associated with the practice of science.

Though after the war the paper was reissued twice, it is not easily obtainable. It was suggested therefore that it should be published again. Then this suggestion evolved into a project for a book which, besides the aforesaid paper published in English, would also contain an article by another Polish sociologist, Florian Znaniecki. This latter paper is even more interesting from the historical point of view because it was published ten years earlier. Moreover, it is longer than Ossowskis' paper and was never printed in English. Even more recently it was decided to add certain post-war texts by other Polish authors, as they would throw light on the further development of the ideas outlined in both former papers.

It was quite accidental that Professor Znaniecki's text had not been published in English. Znaniecki spent the second half of his life in the United States, where he lectured at the University of Illinois (Urbana) and published his works in English. He is still well remembered and often quoted by sociologists.

The other eight texts included in this book have been written in the post-war years. The science of science has been a subject of great interest both in Poland and elsewhere. The texts illuminate certain ideas in this domain, though, of course, they do not form a representative collection. It is hardly possible to deal fully with such broad and varied topics in a publication of this small size. We have only been able to present some pieces of this rich mosaic of subjects. Selection, however, is always arbitrary and therefore 'unfair'. The book does not contain texts presenting some narrow aspects — as a rule more profound and more formally treated. Very recent texts are also omitted. This is convenient for editiorial reasons and seemed permissible because of the historical character of the book. We trust that the texts included in the book will arouse readers' interest in these problems, which are of permanent importance, unless the value of science itself is questioned—which seems hardly likely.

The fact that all the texts have been written by the authors of one nationality is hardly significant. There is little local colour about them; it is possibly perceptible only in those passages which are concerned with the premises that underlie science policy. There is no Polish school of the science of science in the sense of the 'Polish school of logic' (the Warsaw-Lwów school, associated with such scholars as Łukaszewicz, Twardowski, Leśniewski, Tarski, well-known to all logicians). It simply happens that in Poland a fair amount of attention is paid to the problems of the science of science, and since its scope is wide, the approach must be equally wide. For example, the quarterly *Zagadnienia Naukoznawstwa*, published in Polish, began to appear in 1946, i.e. long before similar journals in many other countries.

The articles have been arranged in chronological order, according to their date of publication.

Two main categories of problems can clearly be singled out in this book: methodological problems connected with problem-solving and with correct reasoning in science, and organizational problems connected with achieving optimal structures of scientific institutions, and securing the offectiveness of scientific activities.

Bohdan Walentynowicz

Florian Znaniecki

THE SUBJECT MATTER AND TASKS OF THE SCIENCE OF KNOWLEDGE*

Translated by Christopher Kasparek

Historical roots of the theory of knowledge

Although theoretical reflection on knowledge — which arose as early as Heraclitus and the Eleatics, stretches in an unbroken line through the history of human thought to the present day — nevertheless the most recent times have introduced into these reflections so many new questions and viewpoints so divergent from the earlier ones that we may safely say that we are now witnessing the creation of a new *science of knowledge* whose relationship to the old inquiries may be compared with the relationship of modern physics and chemistry to the 'natural philosophy' that preceded them, or of contemporary sociology to the 'political philosophy' of antiquity and the Renaissance. To be sure, we are still dealing with an accumulation of miscellaneous observations rather than with a systematically and consciously developed scientific whole, but gradually an order is emerging from this chaos and there is beginning to take shape a concept of a single, general theory of knowledge as a separate branch of human culture, endowed with special empirical properties and permitting of empirical study. This theory is beginning to take its place beside such sciences as economics and linguistics as it assumes the traits of a positive, comparative, generalizing and elucidating science. Thereby, too, it is coming to be distinguished clearly from epistemology, from

* First published as 'Przedmiot i zadania nauki o wiedzy' in Polish journal *Nauka Polska*, Vol. IV (1923), No. 1.

normative logic and from a strictly descriptive history of knowledge. The distinction, whose guidelines we will briefly set down here, is not the result of some arbitrary *a priori* designation of the boundaries between the respective fields of human thought, but has developed spontaneously through the emergence — within each of the earlier types of reflection upon knowledge — of problems that have resisted accommodation within its traditional sphere. These problems, gradually concentrating on a common ground outside the scope of purely epistemological, logical or historical thought, constitute one of the main sources of the new science of knowledge.

Epistemology, which may be characterized as a metaphysics of knowledge, seeks to plumb the absolute essence of cognition in general, its limits, and the general conditions of its validity. Thus, for the epistemologist, the essence of knowledge may be defined absolutely and once and for all; his view of cognition is, in his own eyes, independent of the historical evolution of human knowledge. A classic example here is Kant's *Critique of Pure Reason*, [1] which likewise contains implicitly a view of the meaning that research into the actual development of knowledge possesses for the epistemologist; that research presents itself simply as a record of the gradual realization in human history of that absolute essence of cognition that the epistemologist defines independently of history. However, in the past several decades there have appeared epistemological trends in which the earlier theories on the essence of knowledge are represented as being conditioned by the actual state of the disciplines in the respective periods, and thus as historically relative. These trends, while themselves unable to do without general absolute propositions concerning the essence of knowledge (inasmuch as they are still epistemological), nevertheless strive to reduce those propositions to the smallest possible number of formal principles and envisage the main task of the theory of knowledge as the empirical study of the actual development of science based on concrete historical examples. That is the trend, for example, of Schiller's humanism, of the pragmatism of Dewey, Mead and A. W. Moore; and of Rauh's extreme empiricism; and it is approximated by Mach's empiriocriticism and

Poincaré's partial relativism. In short, in these currents, the centre-of-gravity of cognitive theory shifts from metaphysics to an empirical science of knowledge; and precisely to that extent these currents cease to fit within the boundaries of general philosophical epistemology and contribute to the building of a 'science of science' as a special humanistic discipline.

In referring to this evolution within epistemology itself, we scarcely wish to suggest that the latter loses all and any meaning it may have in our overall thinking, and that the empirical 'science of knowledge' can supplant it entirely. The question of the nature, limits, and validity of cognition remains a momentous problem of human thought, and a positive science of knowledge is in no position to resolve it. Like every special discipline, it must leave ultimate problems of this kind to general philosophy; it simply accepts the existence of knowledge as a given collection of facts, being unable as it is to delve into the transcendental nature of the entire field. Its limitation results from — among other things — the fact that it is itself a discipline, just like the disciplines that it studies, and its own limits and validity remain, for the philosopher, as problematical as the limits and validity of any other discipline. The far-reaching significance of the newer epistemological trends in the history of reflection upon knowledge rests mainly on the fact that the epistemologists themselves have begun to realize the necessity of *complementing the metaphysics* of knowledge with an empirical science of knowledge — a necessity arising from, among other things, the fact that the very metaphysics of knowledge expresses itself in historically given particular epistemological theories whose sources, character, and influence lend themselves in some measure to scholarly explanation in accordance with the tasks and methods of scientific investigation prevalent at the time, the personality and social standing of their authors, and so on.

Logic, in company with methodology, strives to create an ideal of perfect knowledge and to point the ways to that ideal. If we have described epistemology as a metaphysics of cognition, we may characterize logic and methodology as the *axiology* of cognition. Considered abstractly, its chief task is independent of the results of studies of historically given scientific and philosophical theories, since what it attempts to do is precisely to impose its ideal on any and all theories and it appraises them accordingly as they approximate to that ideal. An empirical theory of knowledge, in its structure, must submit to some logical ideal. Yet, in the more recent evolution of logic and methodology, a question has also arisen concerning the historical relativity of the logical ideals themselves. Schiller, in his works (among others, in his *Formal Logic*), pointedly emphasizes the dependence of tests of logical perfection on the questions actually posed by science in various fields, periods and environments. American pragmatism attempts to accentuate this same dependence, working from the principle of the dependence of the formal process of cognitive thinking on the practical everyday problems that are thrust upon that thinking for their solution. Even in denying this principle we cannot shut our eyes to the fact that the diversity of actual processes in which cognitive questions are posed and solved does not submit to confinement in any logical system. Similarly, in France, in lieu of the old deduction of methodological rules from general principles of formal logic, we find a tremendous preponderance of monographic inductive studies on methods actually employed in various disciplines and branches of disciplines — without regard to the question of the possibility of their subordination to common ideals of formal perfection. In periodicals such as *Revue de Métaphysique et de Morale*, *Journal of Philosophy and Psychology and Scientific Method* these trends have found appropriate expression. In short, logic, just as epistemology is beginning to demand an ever more empirical theory of knowledge as its complement, and many logicians and methodologists have already made important contributions to this theory.

The descriptive history of knowledge quite some time ago had already produced questions which resisted a convenient fitting within its own framework, questions which might generally be described as being concerned with regularities in scientific development. In Comte's classification of the sciences, the logical order of which is supposed to correspond to the sciences' genetic order; in the attempts at a synthesis of the history of philosophy by Hegel's school, in the works of Ernest Naville, in the well-known history of philosophy textbook by Janet and Séailles based on the concept of the development of philosophical questions, in numerous studies of the histories of individual disciplines (e.g., in *Dzieje myśli (History of Thought)*, part VI of *Poradnik dla samouków (The Autodidact's Adviser)*, Warsaw, 1907–1911), treating the history of a given discipline as the gradual production or emergence of its systematic structure — in these and many similar deliberations we find the beginnings of a type of study logically incommensurate with history proper, i.e. with individualizing history. The history of knowledge, considered descriptively, presents itself to us as a history of creative individuals and groups. Of course, when we reach back into the remote past the discovery of creative individuals becomes impossible for lack of documents; still, a given branch of knowledge is always historically individualized at least in regard to the social group, to the people that has fostered it. To be sure, in recent times, in the most developed sciences the role of individuals of great creativity seems to be weakening in the face of the sizeable number of persons working in a field and the spreading expanse of that field; however, this means only that a strictly historical reconstruction of the evolution of science in recent times must take into account a considerably greater number of creative individuals than in the past, and not that history can today pass over the creative individual.

And between the reconstruction of the history of knowledge as the handiwork of concrete individuals and groups, and the study of an abstractly considered, rational development of problems and solutions, the difference is fundamental and obvious. A particular creative act which, logically speaking, is a link in the systematic evolutionary process of science, is connected for the historian with the totality of a creative personality, which after all takes in not only scientific or scholarly functions but also social, political, economic, religious, esthetic and other kinds. Through his creative work the scientist or scholar is in communion with all other creative individuals in the discipline, independently of place and time — and with them only; as a concrete personality he comes in contact almost exclusively with his immediate social surroundings in the given period: among other things, and perhaps predominantly, he is in contact with people and institutions having nothing in common with learning. A historian who wishes to describe creative persons in all their concreteness, against their historical background, cannot at the same time seek for an internal regularity of the development of learning, which becomes manifest only in abstraction from the historical background against which its particular links have arisen. The study of this regularity thus inevitably parts company with historical description in order to beome a separate, comparative theory of knowledge. Perhaps this latter will succeed in discovering interdependences between learning and other spheres of human activity; but that will require first our isolation of scholarly and scientific activity from the primitive concreteness of life and an understanding of the internal unity of learning.

Thus from epistemology, logic and the history of knowledge there are emerging complexes of empirical questions which are clearly beginning to constitute a separate field of study. Nor is that the end of it. This field is also starting to become enriched with empirical studies of knowledge that have come into being thanks to the initiative of two relatively new disciplines whose early generalizations extended over the entire sphere of cultural life. We have in mind here the so-called psychology and sociology of cognition.

It is difficult to draw precisely a historical boundary between the 'psychology of cognition' and logic on the one hand, and between the former and epistemology on the other; for, as we know, these two disciplines have frequently employed psychological assumptions and theories. There is, however, no doubt that over the last decades psychology — having taken shape as a discipline of quite distinct methods — has approached the question of cognitive thought from a new point of view, different from the old epistemological and logical psychologism. We shall merely point out a few of the newer works that illustrate this viewpoint. Classical examples already today are Hobhouse's work *Mind in Evolution* and Baldwin's *Mental Development in the Child and the Race*, and especially the latter's three-volume *Thought and Things, a Study of the Development and Meaning of Thought, or Genetic Logic*, whose subject, in the author's expression, is the '*modus* of psychic function known as cognition' .* Here belong likewise studies by American psychological schools, influenced by pragmatism, of the biological role of cognition as a tool in the loving individual's adjustment to conditions in the environment. Generally speaking, a characteristic feature of the contemporary psychology of cognition is treatment of the cognitive function in its connection with the total biopsychical life of the cognitive individual as an organism endowed with consciousness; that exactly is the new viewpoint which we do not find in the old 'psychologistic' epistemology and logic and which is the special contribution of psychology, as an already developed discipline, to the theory of knowledge.

It is, however, not only in the construction of psychological abstractions that this viewpoint occurs. Earlier still and independently it makes its appearance in the biopsychological analysis of the concrete personalities of creative scientists and scholars, tied in with the question of heredity. Galton *(Hereditary Genius)* and de Candolle *(Histoire des sciences et des savants)* initiated this kind of investigation; de Candolle's work is continued by Ostwald *(Grosse Männer)*. Lombroso and his school, as is known, linked up studies of genius (extended to include creative scholars and scientists) with psychopathology.

Sociological studies of knowledge began, strictly speaking, with Durkheim's school in France. Of course, as early as Comte the connection between cognitive functions on the one hand and the social system and the entire culture of society generally on the other, epitomized in his famous law of three stages, had already endowed knowledge with a sociological significance; still, the concern here is with the dependence of society's life on knowledge rather than vice versa, and consequently we do not find in Comte any attempt at a sociological explanation of knowledge itself. It was only Durkheim, in his paper *De quelques forms primitives de classification*, who explicitly opened up a new path down which he was followed by Lévy-Bruhl *(Les fonctions mentales dans les sociétés inférieures)*, Simmel and others. Apart from these strictly scientific studies, we find still earlier in Germany, mainly under

* It has not been possible to verify the exact quotation as it appears in Baldwin. (Translator).

the influence of the romantic concept of the *Volksgeist* or *Volksseele*, philosophical investigations into the national character of knowledge, which have developed further in Poland and of late have been revived from a more positive standpoint in connection with the question of building up Polish learning (see, for example, articles by Gawroński, Minkiewicz, Kochanowski, Ossowski, Bujak and others in preceding volumes of *Nauka Polska*). Clearly, we have here a question whose scientific solution requires sociological studies of nationality. Of a similar nature are the well-known generalizations of the school of 'historical materialism' (or more precisely, historical economism) concerning the dependence of learning on forms of technical production and on economic and class systems; the establishment of the latter dependence requires thorough sociological studies of social classes and of their significance for culture. Generally speaking, sociology has introduced into theory of knowledge a thesis of the conditioning of cognition by the life of society — a thesis which in Durkheim's school has been expressed in such extreme fashion that that school seems intent on ascribing to scientific truths exclusively the character of social beliefs that draw their objectivity solely from their universal acknowledgment in a society.

It is clear, however, that whatever may have been the contributions heretofore of psychology and sociology to the theory of knowledge, the problems raised by the latter must perforce extend beyond the boundaries of the former. In the process of the fixing of the scope and methods of these two sciences, the audacious claims of their youth are gradually moderating. A psycho-biological interpretation of conscious life can be applied scientifically only to the most elementary phenomena of consciousness — to those that lend themselves in fairly accurate approximation either to experimental studies or to statistical computations. The 'psychology of cognition' in its higher manifestations, i.e., above all in those which constitute science and philosophy, must be something completely different from the theory of the psycho-organic reactions of a living being to influences from the natural environment, inasmuch as here the conscious individual is dealing not with material forces but with ideal categories, and his cognitive activities cannot be conceived in isolation from the entire ideal system of knowledge with which they are associated. And so far as sociology is concerned, we have shown elsewhere (*Wstęp do socjologii*, Poznań, 1922) *(Introduction to Sociology)* that it cannot study all of human culture but must confine itself to its own special field of social phenomena in the strict sense of the word, i.e., in the sense of functions having for their subject individuals and human groups as well as the values created by these functions. Although it was sociologists who initiated the study of collective cognitive beliefs, those studies must be continued outside the limits of sociology.

Thus the theory of knowledge which is being built up can and ought to encompass, among other things, questions in this area that have been raised by psychology and sociology, without however making itself client to those sciences. This hardly means that psychological and sociological facts and theories cannot at times shed light on particular questions in the theory of knowledge; cooperation among various disciplines is always possible and desirable. The point, however, is for the theory of knowledge, like every other discipline, to acquire its own point of view and its own methods, instead of making use of those of other disciplines. This is essential if only because, as we have seen, questions of the theory of knowledge have been approached from at least five different points of view — the epistemological, logical, historical, psychological and sociological. Thus if the study of knowledge remained client to these disciplines, it could not attain the homogeneous character of a systematic discipline but would persist as a chaotic collection of diverse, often mutually incommensurable, questions and views.

The need to isolate and emancipate an empirical theory of knowledge as a separate discipline with its own tasks and method springs, however, not only from theoretical but from practical considerations as well. The theory of knowledge should provide a basis for all practical efforts at inspiring, broadening, organizing and refining scientific work—and the need for it is more and more to be felt. Particularly in Poland the efforts of many years' duration to maintain learning at an appreciable level in spite of political obstacles and, subsequently, following the recovery of independence, the drive for a rapid expansion of learning on a scale far broader than hitherto, have imposed numerous questions whose solution necessitates a much more thorough, strictly scientific investigation of this, perhaps the most momentous, field of cultural life than has been possible up to the present time. The long-time activity of the Academy of Sciences, of the Mianowski Foundation, of learned societies and of many other institutions supporting scientific work supplies plentiful evidence of the practical need for a theory of knowledge, and a number of specific publications of the past twenty-odd years, begun by *Poradnik dla samouków (The Autodidact's Adviser)* and continued at present by *Nauka Polska*, have been giving expression to that need. Of course, the question has hardly been ignored elsewhere; every learned association, every institution for the support of productivity in the field of knowledge, has of necessity had to ponder it. It is possible, however, that nowhere has it been the object of so comprehensive, persevering and concentrated investigations as in our country, and we owe this not only to the exceptional circumstances which we have experienced but also, and above all, to the fact that in these circumstances there have been people who have devoted their whole lives to the question and who have become centres about which the movement in this field could organize. We are the first country to have a separate Department of Science *(Wydzial Nauki)* in its government; perhaps we will also succeed in becoming the first nation to apprehend the theory of learning in its totality and discover an adequate foundation for it.

While a new science is not created by way of methodological deliberations but by way of positive research still in certain crucial periods reflection upon the subject matter, tasks and method of such a science is essential to its further development. It is in just such a period that the theory of knowledge finds itself today. The numerous scattered contributions, of which we have merely mentioned a small part as examples, must be apprehended and organized from a uniform standpoint, if its subsequent development is to proceed consciously, systematically, with a greater than heretofore economy of effort and a greater fertility of results. Such organization must be preceded by abstract programmatic deliberations in order that the theory of knowledge may the more quickly achieve a planned and self-conscious character. And so let us try, if only tentatively and hypothetically, to draw at full length the boundaries of the new science and to indicate guidelines for its conduct. Certainly we realize full well that this kind of introductory outline cannot be in any case a binding program but at best a sketch suggesting certain ideas which may or may not influence the subsequent development of the science: but in any event they ought to provoke discussion and induce scholars to deepen their reflections on their own scientific conduct.

The subject matter of the theory of knowledge

From the empirico-scientific viewpoint 'knowledge' as a subject of research is of course a complex of cultural phenomena simply *given* to the scholar, much as are speech, art, law or any other field of culture. The theoretician of knowledge, much like the linguist, the theoretical economist or the sociologist, and as opposed, e.g., to the logician or the moralist, must take various phenomena entering into the scope of 'knowledge' in various ages and human communities as objectively existing in equal measure. In his capacity as a scholar he has no right to wonder whether the phenomena given to him as the 'knowledge' of a certain age and society are in fact knowledge from the standpoint of the cognitive ideal that he himself recognizes as definitive, whether and in what measure they are not only phenomena but also truths in his view, logically justified by the tests he himself at present accepts. Such elimination, from studies of knowledge, of value judgments upon the phenomena studied may be difficult considering the fact that, in creating or adopting scientific theories, we have grown accustommed to criticizing other persons' statements as well as our own. Equally difficult is it, at first, for a sociologist to abandon his claims to moral or utilitarian judgment of social phenomena, considering that as an active member of social groups he has grown used to applying moral and utilitarian tests to the social life in which he participates. Yet there surely is no need to demonstrate that a theoretician must maintain complete objectivity in relation to his data (which of course is no impediment to taking an axiological postition concerning theories referring to the data) if he wishes to achieve results consistent with present-day standards of scholarship. A linguist studying linguistic phenomena does not distinguish between 'correct' and 'incorrect', between 'good' and 'bad' speech, but leaves any judgments in that way to grammatians and to writers, i.e., to the practical organizers and co-authors of language.

The concern now is with what phenomena the theoretician of knowledge must include in the field of culture that he is studying. Since he cannot apply his own tests of truth to people's views that are given him historically, since in other words he has no right to regard as 'knowledge' in the historico-cultural sense only those cognitive phenomena which in his conviction constitute true knowledge in the logical sense — it is clear that in the choice of phenomena to be studied he must apply the tests of truth recognized by the people who have accepted the given cognitive phenomena as true, that he must regard as 'knowledge' everything that in the periods and communities studied by him has been deemed knowledge. Only in this way will he avoid the danger of substituting for phenomena a conception entertained in advance concerning the nature of the phenomena, of bending the data of historico-cultural experience to his own subjective conviction as to what those data ought to be like. Such conduct on the part of the theoretician of knowledge will accord with the general guidelines of all the humanities, whose basic feature, as distinguished for example from the natural sciences, is the examination of the phenomena under study in the form in which they have been or are actually experienced by people involved practically with them, that is by individuals and groups living in the respective historic periods and places. Similarly the linguist studying a language considers its words and sentences in the sound and meaning that they possess for the individuals and groups speaking that language.

On this general basis we must find a purely formal criterion that, without presupposing anything in advance about the essence of the phenomena comprising 'knowledge' in a historico-cultural, humanistic construction, would still allow for distinguishing these phenomena from among any other cultural (economic, legal, social, linguistic, esthetic, religious, etc.) phenomena, that would allow a judgment in every particular case concerning whether the given phenomenon constitutes a proper subject of study for the theory of knowledge or whether it belongs in the domain of a different humanistic discipline. The first, in a way an external, test by virtue of which some phenomena may from a humanistic viewpoint bc classed in the domain of 'knowledge', is that for at least certain human subjects (individuals or collectives) these phenomena possess the characteristic of '*trueness*' and in that respect are opposed in their conviction to other phenomena regarded as '*false*'. This conviction manifests itself empirically in the intensional utterances of these human subjects or in the use which they make of the phenomena in question. Any phenomenon to which anyone at any time ascribes the quality of trueness as opposed to the falseness of another phenomenon, we term a *cognitive value*. This trueness or falseness (in the humanistic sense stipulated above) is attributable to cognitive values with respect to something else which occurs tn the role of an object of knowledge, that is of an object to which ihe meaning of the cognitive values refers.

In defining cognitive value in the above fashion, we purposely leave its own content undefined. For it turns out, in examination of materials which history and ethnography furnish to the theory of knowledge, that in various times and human communities various categories of phenomena have been regarded as being capable of possessing the characteristic of trueness or falseness. And since in studies of knowledge we must abide by the guideline of the humanities and take phenomena in the way in which they present themselves to people in the epochs and environments studied, then we too must regard as real cognitive values the categories of phenomena that those people have treated as cognitive values. And so at the dawn of knowledge various words or other symbols have sometimes been regarded as true or false; trueness in the belief of peoples and individuals at lower levels of culture often consists in applying to an object of knowledge the symbol (word or sign) that has traditionally been applied to such objects, falseness — in applying a different symbol. Hence, for example, the innumerable disputes over words in the history of learning.

Futhermore, as true or false may be taken not only single symbols but also combinations of symbols: sentences, arrangements of signs. This understanding manifests itself, e.g., in the efforts aimed at grammatical precision of statements and verbal definitions that so conspicuously characterize medieval knowledge. Finally, truth or falsity is sometimes ascribed not to sense phenomena but to ideal objects and to their combinations — to concepts, logical judgments, conclusions, theories. These objects are real from a humanistic point of view exactly to the extent to which people actually operate with them as being real, as manifested in the words or sensory signs being treated as pure symbols denoting those ideal objects, and those being regarded as true or false with reference to the objects which by their aid we come to know. We find this category of cognitive values, e.g., in modern physico-mathematical knowledge. The theoretician of knowledge has, of course, no right to regard any one of these categories of cognitive values as the sole justified one and to pass over others as not belonging in the domain of knowledge; nor may he substitute one category for another. His research will be fruitful only if in every case he is fully conscious of what kind of cognitive values he is dealing with. His task is not criticism but a thorough plumbing of the content, significance and relationships of the values that he finds in the history of intellectual culture, comparing, generalizing, classifying and explaining them as positively given and actually existing phenomena.

But the object of study in the theory of knowledge is not restricted to cognitive values. These, like all other values, owe their genesis, their historical evolution, their relationships and the role that they play in cultural life to appropriate human operations. These *cognitive operations*, as we may call them, constitute a special category of cultural activities of man's, much as do, for example, social, economic, legal or artistic activities. And although, in contradistinction for example to technical operations, they do not always manifest themselves in material movements, that is, they are above all 'mental' operations, still there is surely no need to prove their existence. For we experience those operations through their very execution; our every act of observation, of definition, of reasoning brings home to us their empirical character. And scholarly investigation can reconstitute and define every operation of another person's from its manifestations in the sphere of cognitive values. For cognitive values are objects of cognitive operations in the capacity of their materials, tools or products; every change in cognitive values has its source in cognitive operations.

Appropriately to these objective manifestations we may divide cognitive operations into three classes. In the first class we will number operations that refer existing cognitive values to the objects to which they apply. Here belongs, e.g., the search for examples to illustrate the meaning of a scientific term, the carrying out of experiments to demonstrate theories, the application of the results of scientific research to everyday practical affairs. We might term operations of this class, operations of *cognitive experience*. They result in a cognitive orientation in the chaos of our data, in the introduction into our empirical world of a certain rational order consistent with the meaning and arrangement of our cognitive values (words, sentences, concepts, judgments, etc.) and conspicuous for the most part in the manner in which phenomena present themselves to us henceforth, and sometimes (in a practical application of knowledge) in the manner in which we attempt to materially transform these phenomena. In operations of this kind, cognitive values play the role of tools with which the cognitive arrangement of phenomena is brought about.

A second class of cognitive operations comprises those which form new cognitive values or modify old ones with the aid of material supplied by observation of cognitive objects. These operations are the best known ones, since it is chiefly they that have up to now been the focus of interest for various portions of the theory of knowledge that has been taking shape. As examples may be cited: the creation of a term to designate a newly isolated class of objects; the suggestion and proof of a new hypothesis on the basis of observed facts; the critique of an old hypothesis by new experiments whose purpose is to test its validity. We would term these operations, operations of *cognitive idealization*, for they create from materials isolated from the chaotic becoming of our empirical world, objects of a distinct type elevated above this becoming, playing, as we have already seen above, the role of regulators by whose help the chaos of empirical data may in some measure be ordered, rationalized. This idealization of course is not absolute since in fact, as the history of knowledge proves, cognitive values in their capacity as components of human culture themselves enter into the general current of becoming, are historically conditioned and mutable; in any event, however, the concern is only with their relative ideality as compared with the phenomena to which they refer.

The cognitive operations of the third class, which we shall term operations of *cognitive systematization*, work with cognitive values in abstraction from the objects to which the latter refer, associate them with each other and systematize them, in the process making use of preexisting cognitive values as material for the creation or modification of others. Here belongs, e.g., the linking up of sentences into the continuous course of an exposition, the deduction of a conclusion from given premises, the elaboration of a general theory to which are subordinated a series of preexisting particular theories, the logical criticism of a theory on the basis of its agreement or disagreement with other theories or on account of its internal content. In saying that these operations work with cognitive values in abstraction from the objects of the latter, we understand by this that they are not immediately involved in the representation, production or modification of the connection between cognitive values and their objects; but this connection is, of course, potentially implied as established in every act of associating cognitive values, or otherwise those values would be devoid of meaning. Much as in speaking we need not refer every expression to an object symbolized by it but employ the established sense of the expressions as something that is empirically associated with their sound and potentially indicates the designated object, so also in systematizing concepts and theories we know what they designate without having every time to illustrate them or to verify them by reference to experience.

Still, it is understood that the foregoing three kinds of operations in the actual course of cognition seldom occur in isolation. Thus, for example, the practical application of a theory often leads to its critical modification, the creation of new concepts is linked to the contemplation of old ones, the systematization of scientific values is sometimes accompanied by their verification and, conversely, the founding of a new theory on experience often takes place parallel to its systematic association with other theories.

In studying cognitive operations, just as with cognitive values, we must maintain the humanistic viewpoint, i.e., regard as cognitive those operations which are entitled to that characterization in the conviction of the acting or 'thinking' subjects themselves. From this it follows, among other things, that the known distinction among logical operations is to the theoretician of knowledge completely unfounded. All cognitive operations are logical in the sense that they possess, in the belief of the cognitive subject, an objective validity, in the sense that they refer to cognitive values as to their objects, actualize them in application to experience, create them, transform them and associate them. To be sure, their validity, considered comparatively, turns out to be only relative; no cognitive operation actually carried out is an absolute cognition; each may be complemented or supplanted by another operation. The theoretician of knowledge cannot, however, on that ground deny them a logical character since, standing as he is on a humanistic position, he is studying not absolute cognition in itself but historically given cognition as it presents itself to historically living people.

On the other hand, all cognitive operations may be called psychological, precisely in the sense that they are carried out by empirical individuals or by human communities and so are subject to psychological study just like any other kinds of operations. This psychological character of theirs does not, of course, mean that they are biopsychical processes; they possess a subjective-ideal character, as opposed to the objective-real character of the processes studied by the biological psychologist. The same may, however, be said generally of any conscious operation; the 'psychology' that must be practiced by the humanistic investigator, the psychology of art, religion or societal life, in spite of the common term actually has few viewpoints in common with the experimental psychology of the biological schools.

The most important source for the distinction between the logical and psychological processes in the domain of cognition actually has been the evaluation of cognitive operations from the standpoint of the standard of scientific or scholarly knowledge that at a certain stage in the evolution of knowledge has been regarded as supreme. Operations conforming to accepted criteria of scientific excellence have been opposed, as being logical, to operations that failed to satisfy those criteria, as being psychological. Thus, for example, syllogistic reasoning in accordance with norms has been distinguished, as being logical, from that which was not so in accordance and which, being 'illogical', has been treated not as a cognitive process but as psychological association. It is obvious, however, that this kind of distinction cannot be maintained in science, which deals with an unlimited diversity of historically manifested cognitive operations and cannot recognize any one of the tests of scholarly adequacy as final and absolute. A cognitively imperfect operation, one qualified as 'illogical' from the standpoint of one test, may present itself as logically valid from the standpoint of other tests; the opposites, known to history, of apriorism and empiricism, of rationalism and mysticism, supply good examples of such a difference in stance.

One might at most consider whether a given operation is cognitively adequate or not from the standpoint of its own criteria to which it seeks to conform, whether it is, in other words, 'logical' or 'illogical' subjectively. But even then, on close examination, it transpires that such a classification is pointless since in fact there are no perfect cognitive operations, even from the standpoint of their own criteria. For these criteria constitute ideal models which cognitive operations attempt with lesser or greater success to realize in the values they produce, but which they never realize completely in the face of the difficulties that cognitive thought encounters from the reality that is the object of cognitive values. An operation working with cognitive values which from the standpoint of certain criteria are appraised as false by other operations, is still a cognitive operation just as is one whose values under the provisions of the same criteria are recognized as true, since the operation would not be consummated did it not lay claim to the truth of its materials, tools or results; and the differences in validity of the claims of particular operations are not absolute but relative, conditional not only on the nature of the criteria but also on the degree of approximation to their realization. Likewise, e.g., technical operations producing objects defective or useless by their own technical criteria, are still technical operations just as are those whose products most nearly approach the acknowledged ideal of excellence.

The analysis, description, and classification of cognitive phenomena

It is obvious that the theory of knowledge cannot satisfy itself with a simple description of individual cognitive values and operations in their concrete relationships — that is the business of the history of knowledge. The task of the theory of cognitive phenomena, as of any theory of a scholarly character, is firstly the analysis and classification of those phenomena, and secondly the causal elucidation of facts occurring in the field.

In scientific analysis and classification the concern is above all with the isolation, from the concrete complexity of the phenomenal systems given to us in observation, of relatively simple elements and arrangements which can be discovered in various combinations and repeated in variable circumstances; and subsequently, with the systematic designation of basic similarities and differences among these elements and arrangements that might permit us to divide them into classes hierarchically ordered as to their generality.

Relatively the furthest advanced in the literature so far is the scientific analysis of cognitive phenomena. Of course, one might register numerous reservations as to the methodological premises of many works in the field; one might wish for a greater standardization of methods, a more conscious cooperation, above all a broadening of analytical studies to various, as yet almost untouched, sets of cognitive phenomena and a strict separation of exclusively theoretical studies of knowledge from attempts at a normative critique, evaluation and reform of knowledge. These are, however, questions whose consideration would require far more space than we presently have at our disposal; and other questions need more urgent attention.

Specifically, for example, the present state of the theory of knowledge is much more unsatisfactory in the classification of cognitive phenomena than it is in the domain of analysis. True, we do find on the one hand various classifications of cognitive operations in logic, epistemology and the psychology of cognition, and on the other hand the so-called 'classification of the sciences' to which Comte first strove to give a positive-scientific character: still, all these attempts have one flaw in common. Of the entire enormous wealth of cognitive phenomena that we find in the world of culture, they consider only a small part, most often selecting and systematizing it on the basis of criteria adopted *a priori* and associated with an ideal of cognition approved of by the given author. A truly scientific classification, however, ought to be completely objective and impartial, based on criteria deduced *a posteriori* from the empirical properties of the phenomena studied and embracing all the known phenomena in the field; moreover, it ought to be not only a framework for the systematization of results attained but a tool for discovering new, as yet unobserved phenomena and new properties of phenomena already observed. The only kind of classification that can satisfy all these requirements is a *genetic* classification. The question is, however, whether such a classification is possible in the theory of knowledge and on what premises it should rest. To be sure, Comte already held that this classification of the sciences was at once logical and genetic; for reasons mentioned above, however, we cannot accept that assertion as definitive.

In considering the above matter we will have to limit ourselves here to the question of the classification of cognitive operations as being less complicated than the classification of values. For every cognitive value is first and foremost the product of a certain operation; consequently it is determined genetically above all by the nature of the latter, and beyond that depends on the kind of materials and the tools which the operation employed. We can express this still another way in traditional terms. Every cognitive value, every 'truth' presents itself as the solution to a certain theoretical problem. The posing of the problem depends solely on the cognitive operation; its solution, by contrast, is conditioned additionally by the data and investigative tools which the operation has at its disposal. This is expressed at its most elementary when the problem posed requires a quantitative solution; thus, for example, the problem of the size of the sun or of the earth's distance from the sun has been subject to many different solutions depending not only on the logical differentiation of the actual mode of thinking but also on the data on which astronomy has relied and on the instruments that it has employed. The number of possible solutions reduces to two, affirmation and negation, if the problem is of the kind such that its solution depends entirely on the existence or non-existence within precisely defined limits of a certain hypothetical, just as precisely defined, datum; this is an extreme case, possible only in the realm of certain theoretical conventions.

Thus the diversity of cognitive values as products is greater than the diversity of the operations producing them. Moreover, though, a value already produced can serve further as material or continues to differentiate. Let us recall, for instance, the various changes of meaning that have affected the atomic theory, concepts of psychological association, and the assertion that the invention of the steam engine was the cause of contemporary capitalism — all depending on the questions for whose solution the above cognitive values have been used. The classification of cognitive values cannot, therefore, take them in all their historical concreteness but must define them in relation to the specific determinations to which they are subjected by certain kinds of cognitive operations. The latter are, to be sure, accessible to historical investigation only intermediately through the changes that they bring about in the domain of cognitive values; however, it is easy to obtain an exact definition of every operation by comparing its various manifestations under various circumstances.

If following these preliminary remarks we consider the historical evolution of human knowledge, it will present itself to us as the creative growth of cognitive values and the emergence of new cognitive operations. Clearly, the empirical theory of knowledge need not and cannot ponder a question belonging to the metaphysics of knowledge — namely, whether and to what extent scientific creativity is in an absolute sense the creation of objectively new truths or merely the discovery of subjectively new truths which have existed outside the realm of human consciousness always, or rather timelessly. For in any case the act of introducing, into the empirical domain of human cognitive values, a truth that had not previously been in that domain constitutes, empirically, an original and creative act, just as does the creation of a new work of art unlike any existing up to that time, or of a new technical invention.

This creative development, precisely to the extent that it is creative, is impossible to *elucidate scientifically*. For scientific elucidation as applied to phenomena is basically causal elucidation, i.e., the indication of a necessary *real* relationship among *facts*; thereby does it differ from *explanation*, which expresses an *ideal* relationship among the elements of a logical system, and from *philosophical* elucidation, which seeks to deduce secondary properties of phenomena from their basic *essence*. Creativity eludes the principle of causality and must in some way be excluded from scientific elucidation, as we shall see later. The science of knowledge may, however, systematically describe the latter in its creative development, basing on the genetic association of cognitive operations their essential definitions and division into classes.

For every new cognitive operation is associated with some preceding operation, is a result of the latter's differentiation, is a new variant of it, or arises as a new intellectual function in a logical synthesis with other operations. Thus, to take an example, the question of the relationship of the culture of human societies to their racial composition arose as a modification of the old problem of the relationship of the 'soul' to the 'body', as applied to a community instead of to the individual; the new problems in Newton's physics sprang from problems of Galileo's, Kepler's and Huygens' in a synthetic union; the new components of Kant's philosophy arose genetically partly as formal modifications, partly as functional differentiations of certain philosophical concepts of Hume's, Wolff's, Rousseau's and others' and of certain mathematical, physical and ethical questions. Every new cognitive operation is thus logically and genetically a continuation of cognitive operations previously carried out. We may call this assumption, indispensable to the scientific treatment of knowledge in its historic becoming, the *principle of continuity in the evolution of knowledge.*

In accepting this principle, the theory of knowledge must nevertheless beware of two pitfalls: historical empiricism on the one hand, and teleological rationalism on the other. Although every new cognitive operation harks back genetically to previous ones, this hardly means that continuity in the evolution of knowledge rests on the historic continuity and cultural homogeneity of the society nurturing the field of knowledge. New cognitive problems do not necessarily arise in connection with questions that have concerned scholars working in the society in the immediately preceding period of its history. They can constitute a continuation of questions posed and solved a thousand years earlier and in entirely different social groups characterized by fundamentally different cultural complexes. The genetic connection between Renaissance thought and ancient Greek thought, and the taking up by Japanese scholars of scientific research in the mode of European and American scientists, here supply obvious examples. We shall see later in what sense and in what measure one may speak of the relationship of learning to factors of a societal nature; that relationship, however, is in no event one of such a kind that the creative evolution of learning might take place solely within the historic continuity of human communities and constitute merely a part of the general evolution of the cultures of those communities.

Similarly false would be the assumption, typical if different examples of which we find in Comte and Hegel, that there exists in the evolution of knowledge an immanent rational order, definable once and for all from the standpoint of the requirements of a scientific final cause; that this evolution is essentially a gradual, objectively necessary realization of some perfect type or system of cognition and that its every step constitutes a logically indispensable link in some universal regularity. On the contrary. The actual evolution of knowledge at every step displays features of irrationality and unpredictability in the generation of new evolutionary lines. Every cognitive operation is, to be sure, logically a continuation of some preceding operation, but there is no objective necessity, real or ideal, for just this new operation and no other to be consummated in connection with the preceding operation, or even for just these operations and not others already carried out to find continuation at all in new operations. The principle of continuity in the evolution of knowledge has to be complemented by a principle of the *freedom and unpredictability* of that evolution. Having some cognitive operations already effected, we cannot aver that any particular designated new operation had actually or logically to arise from them. Precisely to the extent that the new operation is original and creative, its accomplishment cannot be subject either to actual or to logical determination.

In studying the history of knowledge, we repeatedly find that some cognitive operations, in defiance of historical probability and the requirements of final cause, do not find a logical continuation, that an opened evolutionary line breaks off for many years or even forever. Thus, for example, the heliocentric hypothesis of Aristarchus of Samos languished for centuries and had to be recreated anew by Copernicus; the Cartesian theory of vortices and the *characteristica universalis* of Leibniz have not found a logical continuation and, despite recent attempts at their resurrection, it is uncertain whether the respective evolutionary lines ever will have any continuation. And conversely, new cognitive operations sometimes turn out to recur not to some established evolutionary line which, rationally considered, 'ought to' lead further, but either break with the immediate tradition and, to initiate a new line of evolution, reach back to a remote, forgotten past, or search for new points of departure outside the field to which we would like logically to restrict them, e.g., in some other discipline or even in a cognitive activity attendant to practical life and hitherto unconfined by scientific method. And so, e.g., certain trends in contemporary philosophy have made recourse to medieval problems which modern thought is accustommed to regard as sterile, incapable of further development; numerous currents that have arisen in the past century in the humanities have had their sources not in the past histories of those disciplines but in new theories and methods of the natural sciences; we find the genesis of American pragmatism first and foremost in cognitive reflection on practical questions — of technology, economics, religion.

Furthermore, even when the point of departure for subsequent evolution is an operation or set of operations whose prolongation presents itself to us as in fact probable and requisite to the rational progress of learning — this subsequent evolution can still go in a completely arbitrary direction, even one totally inconsistent with our anticipations. Looking into the future, we should realize that there may be various ways of logically extending a certain operation, that various new operations can recur to the latter and that there is no determination, binding in advance, on whose strength a future operation that will recur logically to the operation presently under consideration, would have to be just so and not otherwise. The genetico-logical association between operations comes into being only at the moment of execution of the respective operations; it is not limned in advance prior to their appearance in the sphere of cognition. A genetic relationship is negative, not positive; this means that a new operation could not appear if there were not already another operation of which it is logically a continuation, hence that the new operation must stem from a designated operation, rest on it for its actualization; but an operation already existing in the realm of knowledge does not require the actualization of designated new operations. There could have been no Newtonian physics without Galileo, Kepler, Huygens; but Galileo, Kepler and Huygens did not have to be followed by Newton and his problems; their theories might have gone undeveloped, or a savant might have happened along who would have rested other problems on them, today perhaps altogether impossible to conceive. [2]

Undeniably the scale of possibilities for the future evolution of a science differentiates into various levels of probability. That differentiation has its source in the real and ideal limitations that human creativity sets for itself. The real limitations stem from the fact that cognitive thought, like all human activity, is in lesser or greater degree dependent on the circumstances with which it deals, and as a result is in fact capable of realizing only some of its theoretical possibilities; in just the measure to which the dependence has come into effect (and we see an example of this in common mental habits), the possibilities realized will be precisely those to whose realization the given circumstances are most conducive. And the ideal limitations of future possibilities stem from the voluntary conformity of scientific thought to certain guidelines marked out by its previous development. For, while the growth of knowledge is not originally and spontaneously purposeful, still at higher levels of scientific culture attempts at introducing a teleological element into it multiply. We find in history numerous attempts at a planned organization of future scientific work, attempts based for the most part on results and criteria in each case already achieved. The programmes of scientific and philosophical schools, and normative works in logic, methodology and the classification of sciences supply guidelines of this kind to which, within more or less broad limits, cognitive creativity consciously conforms. In spite of that, as we know, sometimes in lieu of the cognitive operations for which the most conducive cultural conditions are ready, there unexpectedly appear other operations for which reality up to that time provides no support, and in face of all external limitations they create for themselves, despite difficulty, the indispensable scientific situation; occasionally, too, the best-laid plans for the organization of scientific work are disregarded or are altered by some new and unforeseen individual initiative.

This relatively unpredictable character to the development of science, however, hardly impedes the descriptive systematization of cognitive operations, provided only that the systematization does not attempt immediately, working from aprioristic categories, to be exhaustive and to possess a perfect logical form. However original a certain cognitive operation may be, it is never absolutely unique in its kind, if only because it is never an absolute beginning. Cognitive operations can be similar to one another in various degrees, and that makes their classification possible. And the history of knowledge demonstrates that the number and diversity of cognitive operations has grown enormously, even in the relatively short period of civilized life that we know from historic documents; evidence of this is, among other things, the emergence of numerous new disciplines and branches of disciplines. On the strength of the principle of continuity in evolution formulated above it may be concluded that, notwithstanding many cases of the breaking off and inhibition of evolutionary lines, the history of knowledge may be represented as the gradual emergence of a whole immense wealth of cognitive operations from one or several original operations through the progressive differentiation of forms and functions originally undifferentiated.

We thus find here a basis for a genetic classification of the elements of cognitive thought. Much like the biologist in the genetic classification of living organisms, the theoretician of knowledge must assume that operations arising by differentiation from a common source resemble each other more closely in a formal (or a functional) regard than do operations derived from diverse sources, and that this similarity among them is the greater the closer is their common genesis, the fewer the intermediate links between them and the more primitive operation in which the historic lines leading to the operations under study have had a common beginning. It is easy to realize how by this method, all the operations that we shall discover in the history of cognitive thought, can gradually be defined and brought into a classification system based now, not on arbitrary criteria drawn from one or another logical, epistemological or psychological theory, but on the actual bonds of genetic kinship. Naturally, such a system will be far more complicated than is, say, a Kantian classification; and further complications will arise when we pass over from elementary cognitive operations to the characterization of systems of operations — and, finally, of those great complexes of cognitive operations and values that we call 'disciplines'; in fact the main charge that must be lodged against all the attempts heretofore made at a general theoretical treatment of cognitive phenomena is that of illegitimate, we might ever say naïve, simplification of the object studied, an insufficient awareness of the gigantic wealth and complexity of this field of culture that we call knowledge.

It should further be stressed here that a genetico-descriptive systematics of cognitive phenomena is by no means merely a scholarly re-creation of the cognitive *past*. With the emergence of new, genetically later cognitive operations, the old, more primitive ones do not at all atrophy. They persist and continue to be carried out with only slight modifications of their original form. Likewise old types of cognitive values continue to recur in communities of the highest cultural standing and in the intellectually most highly developed individuals. There is no discipline in which there do not in fact coexist cognitive operations and values corresponding to various historical stages in the evolution of knowledge. Stages whose genesis lies in prehistory are represented in all their vitality to this day in the thinking of so-called savage peoples and among the uneducated classes of civilized societies. They manifest themselves even in individuals familiar with the newest criteria of contemporary learning, in respect to questions of everyday life, both in so-called common sense and in numerous uncritical so-called superstitions. Finally, the most primitive, the most elementary cognitive operations and values are probably to be found in the thinking of small children and perhaps also in that of mentally subnormal individuals. The genetico-descriptive treatment of cognitive phenomena should, certainly, rest mainly on the history of knowledge, since it is there that cognitive phenomena stand out most particularly and clearly in documented forms and symbols and it is there that the emergence of truly, objectively new phenomena is best traced; nevertheless, the domain of those studies spreads likewise over the entire field of contemporary intellectual life, not only to its peaks. For here we can observe the content, meaning and changes in the cognitive values of others and even evoke cognitive operations experimentally. Comparison of currently experienced contemporary cognitive phenomena with those that we recreate on the basis of historically available materials should become equally fruitful a method in the theory of knowledge as is the comparison of currently observed and artificially produced biological processes with the data of paleontology, comparative anatomy, animal and plant geography, etc., in genetic studies in biology. [3]

There are, of course, basic differences between the creative emergence of a new cognitive operation for the first time in the history of knowledge and its subsequent reproduction by an individual or social group that in its intellectual evolution has attained the appropriate cognitive level. In the second case the development of the operation is very seldom spontaneous; usually it occurs under the influence of unreflective or of reflective instruction which shortens the evolutionary series, considering only superficially or skipping certain links; and often the instruction strives to alter its direction, leading the individual or group to an idea with which they are unfamiliar along a totally different route from that by which mankind first arrived at it. Thus for instance the order of systematic exposition of a discipline in schools often diverges radically from the evolutionary order of the discipline. Nevertheless, in spite of this we find numerous data permitting at least a hypothetical conclusion that, appearances to the contrary notwithstanding, the same evolutionary operation possesses essentially the same genesis, the same indispensable antecedents in the intellectual development of the nation, class or child as it does in the intellectual development of mankind.

This hardly means that the nation or child must pass through all the cognitive experiences that mankind has gone through, that cognitive 'ontogeny' must be an exact, if abbreviated, recapitulation of the cognitive 'phylogeny'; to the genesis of a certain cognitive operation, the countless past deviations from the course that has led to that operation, the disrupted collaterals, have of course no logico-genetic significance, although they may in fact have been indispensable in one or another set of historical circumstances. The point is only that — as we see clearly in considering the great differences in intellectual levels required for certain operations — a genetically late operation cannot be carried out for the first time by an individual or group assimilating past scientific achievements prior to a genetically earlier operation from which the first operation in its historical evolution has logically derived. One cannot learn differential calculus prior to addition and subtraction nor understand the heliocentric theory prior to at least a schematic appreciation of the geocentric theory from which we begin our reflections on astronomical phenomena, nor grasp Kant's philosophy without at the least an approximate acquaintance with the problems of the sophists and of Plato. If artificial instruction passes over these earlier levels, the nation or person in reality does not reconstruct ideas for whose reconstruction he is unprepared but only repeats words or symbols.

unless — as is generally the case — he makes a spontaneous effort and independently reconstructs the logical evolutionary sequence in his own thinking.

Observations of this kind have, as we know, led to new trends of thought in psychology and pedagogy, according to which the old teaching methods distort the 'normal' evolutionary course and must be supplanted by new ones, in which for example instead of the teaching of geography beginning with a description of the earth, it begins with a description of the school and of the village or town in which the school is located. Similarly, applying an analogous general premise to the intellectual development of backward nations, we ought to conclude that we should above all, emulating the evolutionary order of the past, support the highest types of intellectual creativity and realize the greatest possible cognitive ideals, which will radiate over the generality of society, rather than, as has usually been done up to now, place the main emphasis on the diffusion among the broad masses of superficial presentations of the results of other people's work.

If this principle of *irreversibiliy in the evolutionary order of knowledge*, which we are here positing, does indeed prove correct following deeper and more detailed study, it would open up an extremely important field of practical applications for the theory of knowledge. Here, however, we are approaching a second great category of questions that the new science faces — questions concerning not the systematic description of cognitive phenomena but a causal treatment of the relationships obtaining among them. This treatment must seek, in the creative evolution of new operations and the growth of new values, an element of repeatability on which could be based the formulation of laws, the forecasting of the future and the planned steering of the further evolution of knowledge.

Causal treatment of the relationship of cognitive results to the circumstances of cognition

Every discipline aspires to the greatest possible broadening of the field of its causal explanations, this both from theoretical considerations — inasmuch as the discovery of causal laws permits the subordination of the unlimited multiplicity of repeatable processes to a few general concepts — and from practical reasons, inasmuch as only a causal treatment makes it possible to foresee future phenomena and to direct them. The theory of knowledge must therefore also be concerned, without in any way ignoring the creative character of cognitive evolution, to nevertheless find such access to cognitive phenomena as to explain causally with all possible approximation every particular moment in the enormous system of processes that go to make up the history of human knowledge.

To this end it is essential first of all to isolate, from the cognitive life of mankind, relatively simple processes which might be divided into classes; only then, and considering the processes in each class as being approximately repeatable, will it be possible to apply to them the principle of causality. What we have said before about a genetic classification of cognitive operations, can serve here as a point of departure. If the concern is above all with operations that are merely the reconstitution of preceding ones, than the element of repeatability is already intensionally contained in them. Every cognitive operation — even the most original — at its next repetitions (e.g., in school exercises) is of course subject to a certain regularity which renders the cognitive operation more or less exactly accessible to causal treatment. It is in this very repetition, however, that the theory of knowledge is least interested, important as the repetition is to pedagogy. The more creative, the newer, the more uncommon a certain cognitive operation is, the more it draws the attention of historians and theoreticians of knowledge. Clearly, though, these same traits make its causal treatment difficult. Consequently a compromise of sorts imposes itself.

Between the most original operation and the most passive repetition of another operation exist countless intermediate degrees of originality or repetitiveness. If the precision with which we can treat cognitive processes causally varies directly as their uniformity, then the theory of knowledge must reconcile itself to the fact that the more original and creative a cognitive process, the less precise will be its causal treatment, in other words, the less significance will be possessed in the whole by that element of it which may be subject to the principles of causality. That element, as we shall see, is detectable in every cognitive process; the only question is that of its relative importance. That which, with a suitable method of causal explanation, remains unexplained in the operation of a pupil doing an assignment, possesses little significance in establishing the laws of intellectual life and in foreseeing the future; on the other hand, even the best scientific methods, applied to the creative work of a Copernicus, Newton or Hegel, will have to leave unexplained aspects of scientific activity that possess a first-rate importance for the coming into being of knowledge and for the direction of its future development.

Combining the foregoing deliberations with the question of the genetic classification of cognitive phenomena, we shall say that in every class of these phenomena laws of causal dependence will no doubt be detectable; the more diverse, however, are the phenomena contained in that class, hence, other things being equal, the broader is the scope of applicability of the given law, the smaller a role the law will play in elucidating any particular phenomenon to which we apply it.

Of course, no complete, actually occurring, concrete process of creation of a cognitive value will yield to reduction to any causal law. For its causal treatment would have to consider both the mental act itself and the overall conditions in which the act operates and which co-determine its results. And this actual, concrete combination of a certain operation with certain conditions presents itself from a causal standpoint as incomprehensible and unpredictable, unique in its kind; for any attempts at its explanation would require an inexhaustible causal regression, the consideration of all preceding facts that account for just this act happening upon just these circumstances at just this time and place. Therefore it is necessary to isolate, from this whole complex process, certain basic moments to which causation would be applicable; it is necessary to break it down into relatively simple processes each of which can separately be a link in causal chains.

Such analysis of a process of cognitive creativity is possible and leads to two different and mutually complementary types of questions. In questions of the first kind we will be concerned solely with the relationship of the results of cognitive creativity to its given circumstances, and we will not elucidate the act of creativity itself; while in questions of the second kind we will consider only the causal elucidation of the actualization of a designated cognitive operation, leaving aside its given circumstances and results.

In posing the first kind of questions we must regard a certain category of cognitive operations as being constant. In other words we posit that, in all cases studied and compared, there have been and will be carried out cognitive operations of a certain class, differing among themselves only within the limits set by the definition of the class, and that we need not take into consideration perturbations that may have resulted from not carrying out appropriate operations or from carrying out operations of a different class or such original ones that no generalization is suitable to them. We find further that given the afore-mentioned constancy of a certain category of cognitive operations, there obtains a constant relationship between a certain set of circumstances with which such an operation works and the value produced by the operation. There exists a necessary and sufficient set of circumstances for the realization of a certain cognitive value by a certain cognitive operation. These circumstances are necessary in the sense that until they have been realized, the operation in question cannot create precisely the value that it intends to create, but certain other operations must first be carried out in order that these circumstances may be realized; and they are sufficient in the sense that if they are given and an appropriate operation makes use of them, it will produce the required value. Thus if we find that cognitive operations of a certain class (insofar, of course, as they are carried out at all) produce values of a certain class only when, and whenever, they deal with a certain set of circumstances, then, assuming the operations to be constant, we may conclude that between this set of circumstances and values of the corresponding class there obtains a constant relationship.

The principle formulated here is simply an expression of a conviction of which we make use at every step, in popular reflections on knowledge and in practical activity seeking to stimulate or to restrain scientific creativity in a given direction. Thus for example we are convinced that a chemist who is given a certain stock of previous scientific achievements and a chemical laboratory supplied with materials and instruments, insofar as he wishes and knows how to work scientifically at all, is going to discover new chemical truths, and not astronomical or linguistic ones; what is more, having a precise knowledge of the branch of previous chemical achievements that presently interests him, the kind of materials and instruments that he has at his disposal, and his working technique, we can place predictions about his discoveries within fairly narrow bounds, especially if he is not an exceptional genius but simply a gifted scientist, which is to say, if his cognitive faculties do not extend beyond a certain set range of differentiation.

This relationship between a certain set of circumstances and a cognitive value of a certain class produced under these circumstances, given a constant category of operations, is still not, of course, a causal relationship in the modern sense of the word; for a causal relationship, as it is construed by recent science, is a relationship among processes, or changes, and not a relationship among substances or objects. Nevertheless, on the basis of that relationship there may be realized relationships possessing all the features of a causal relationship. Let us suppose that there occurs a change in cognitive circumstances; inevitably the cognitive value produced will likewise change (always, of course, assuming the constancy of the given category of operations). Let the chemist learn that some previous chemical assertion on which he has been relying has been proved erroneous; let the material or instruments that he employs change: the results of his work will be different than they would have been without the change. Insofar, indeed, as there exists a strict relationship between a given set of cognitive circumstances and a given class of cognitive values produced by their help, we may postulate that a specified kind of change in circumstances will always and everywhere evoke a specified kind of change in the cognitive values produced. And a relationship of this kind among changes corresponds to the concept of a causal relationship.

The task of the theory of knowledge in this realm of questions would therefore be the discovery of laws governing that relationship, i.e., of assertions formulated more or less as follows: wherever and whenever, given a constant relationship obtaining between a set A of circumstances and a class B of cognitive values, there occurs in cognitive circumstances a change of type m, then there must occur a change in cognitive value of type n. The studies in question must of course be conducted by way of comparative observation and (within certain limits) experiment, removing individual elements from systems of cognitive circumstances and studying the results of the changes produced in cognitive values by the changes in those elements, given the demonstrated or postulated immutability of the remaining circumstances. Obviously we cannot expect investigations of this kind soon to attain a complete precision; nevertheless, what kind of significance, theoretical and practical, even their tentative and approximate results can have, will become clear when we have enumerated a few kinds of special questions belonging to the category under discussion here.

(a) *The question of the relationship of the development of particular fields of knowledge to factors introducing previously unavailable materials into the research* (or, conversely, removing certain materials from the research). This is an old and familiar question; much has been said and written on the influence that, e.g., the telescope's broadening of the field of observation has exerted on astronomy, that the consideration of microscopic data has had on biology, that the opening up of new territories — thanks to scientific expeditions — has had on geology or anthropology, that the accumulation or destruction of records has had on history, etc. In spite of that, however, attempts at a scientific treatment of the problem are rare, albeit the very formulation of the problem in terms of an 'influence' (by which is usually understood a causal action) has seemed to call for a more exact posing of the problem. It will not suffice to describe historically the new truths that have been worked out by scholars using the microscopic data or the new discoveries that have been made by this or that geological expedition and the conclusions that have been drawn therefrom; in order to base on these facts any generalizations as to the relationship of biological truths to the microscopic data or of geological truths to the results of the scientific expeditions, it is necessary to discover in them some regularity that warrants the drawing of conclusions about unknown facts on the basis of known ones. As long as we put the question in terms of the creativity of this or that scientist who has discovered these or those truths with the aid of microscopic observation or thanks to materials supplied by a scientific expedition, we have no right to speak of the 'influence' of microscopic observations on scientific theories but only of the 'use' made of these observations by scientists; in other words, we have to formulate the relationship between the microscopic data and scientific theories not in terms of cause but in terms of *final* cause. Yet popular thinking about science actually and rightly does assume the existence of a regularity of a causal nature in facts of this kind, although it does not know how to isolate the regularity.

Isolation of this regularity will become possible as soon as we reduce each of the facts involved to a formula containing: (1) a statement of a specified field of scientific work in whose course materials of a certain kind have served scholars of a certain range of faculties and interests in discovering truths of a certain type; (2) affirmation of the appearance, in the sphere of the research, of a given type of new materials (sometimes the removal, from the sphere of the research, of specified materials that had previously been accessible, e.g. as in the destruction, by some cataclysm, of paleontological relics or historic documents); (3) the determination of a change brought about in the type of truths discovered by the research, hence in the type of cognitive values produced through the research, due to introduction of the materials.

If it turns out that in each case when new materials of the same kind have been made available to researchers, the results of the research undergo changes of the same kind, and supposing that other circumstances have remained basically unaltered, then we shall have a right to regard the change in results as an effect of introducing those materials into the sphere of the given science. So, for instance, let us suppose that we have compared the results of biological research in which microscopic data have been used with the results of otherwise similar research prior to the use of those data; we will probably be able to determine precisely, though of course making allowances for a fairly broad range of possible differences, what kind of specific consequences are brought about by the introduction of microscopic materials into biological research, and foresee with complete certainty, though only again within specified limits of possible differentiations, what will be the result of introducing microscopic observations into biological studies in which they have not yet been taken into account. In the same manner we will probably succeed in generalizing and predicting modifications in the type of scientific results that are produced, e.g., by introducing judicial records into research on a social system, psychological materials into economic questions or technological theory, etc.

The examples cited here are familiar and topical, and we have formulated them rather cursorily; however, it is clear that countless further questions can be posed here and that the answers to those questions will not only have meaning for the elucidation of many facts from the past but will also provide valuable practical hints for the future organization of scientific material collection. One must only realize that we face here questions of an experimental nature requiring numerous comparisons of facts from the early as well as the most recent history of knowledge — of facts accurately described and precisely defined. Deductive explanation or prediction that materials of one kind or another *logically* had to, or logically will have to, lead researchers to one or another kind of results, will not suffice; the actual relationship between the results of research and their material has never been and probably never will be completely in accord with the predictions of logicians.

(b) *The question of the relationship between scientific results and technical instruments.* This question is contiguous with the preceding one since some technical instruments, such as the telescope and microscope, serve mainly to supply observation with materials which otherwise would be unavailable. We shall, however, pass over that aspect of them; at the moment we are interested in something else. A technical instrument not only broadens the range of data (of objects or of their properties) but likewise influences the kind of scientific problems posed and solved in respect to the data. Typical in that respect are physical instruments. The materiality and precision of any technical tool suggests and facilitates problems of a certain kind, while on the other hand diverting the researcher's attention from other problems to whose solution the material tool affords no assistance. Through introduction of the instrument into the cognitive circumstances, the value produced is clearly altered, and the changes seem to be so specific and so extremely characteristic of all facts of this kind that the attainment of positive results in comparative studies of this question ought to be relatively easy for the theoretician of knowledge. We will point, by way of example, to the differences between Aristotle's physics and modern physics, between introspective and social psychology on one hand and laboratory psychology on the other, between the position of the surgeon and the internist in diagnosis and therapy, etc. We shall venture to hypothesize that the result of introducing technical instruments into cognitive circumstances is the introduction of a quantitative element into the cognitive values produced, so that wherever a material tool enters a field of scientific research, there will appear (barring counteracting causalities) a tendency to impart to discovered truths a measurable character — in extreme cases, mathematically calculable. More thorough research will confirm or overthrow this hypothesis, and in any event will uncover further relationships of this kind.

(c) *The relationship of the results of research in particular fields of knowledge to cognitive values serving as tools*, most prominently to so-called heuristic concepts and methodological principles. Various aspects of this question have, as we know, often been discussed. We find at every step, in methodological studies and critiques, deliberations on the scope of applicability, e.g., of causation, of the concepts of matter and energy, etc., and on the results of their application. The objective, however, is to introduce a new viewpoint — experimental-theoretical instead of methodological-normative — into these deliberations. Methodology is generally content to point out the logical consequences stemming from use of this or that principle or concept, drawing therefrom normative conclusions concerning what principles or concepts *ought* to be used in a given area of research in order to achieve a certain ideal type of truth. But the logical consequences to which the use of certain cognitive tools leads in methodological reasoning, are hardly identical with the actual results of their use in an empirical, historically given course of scientific research. Logical analysis of heuristic principles and concepts can only indicate the direction in which research employing these principles or concepts would develop if it made exactly the use, and only that use, of them which is consistent with their meaning as given by the methodologist. In reality, though, that use may be entirely different; the influence exerted by heuristic principles and concepts on cognitive values produced with their help can only be stated *a posteriori*, on the strength of their actual application, since it depends not only on their logical essence as independent cognitive values but also on their historic function as tools of cognitive activity, for which they are merely elements in cognitive situations. Therefore, in order to discover in this influence an actual regularity other than the normative regularity required by methodology, it is necessary to compare changes arising in cognitive values due to the introduction, into the research in question, of one or another heuristic principle or concept — e.g., the principles of the conservation of energy or of entropy into the realm of sociological research, the concept of 'human nature' into economic research, etc. Of course, the actual regularity, as we have already emphasized, likewise does not exhaust that entire aspect of cognitive creativity, but in any case it will throw a new light on it.

(d) *The relationship of the results of cognitive activity to symbols serving to designate cognitive values or their objects.* This question is tied to the preceding one, to the extent that at a certain stage in the evolution of cognition symbols themselves play the role of cognitive values and may serve as tools in the same sense as do heuristic principles and concepts. Nevertheless, in more recent science there prevails the sense of a symbol as a special kind of medium which is not itself a cognitive value but only a distinct type of tool for operating with cognitive values. Appropriate studies in the theory of knowledge would therefore cover the following particular questions: the cognitive influence of specified changes spontaneously occurring in language (e.g., the influence of the atrophy of certain flexions or of the rise of auxiliary words); the consequences of translating from one language into another, in connection with the question of the influence of differences among various languages on scientific creativity; the results of efforts to create an international scientific language; the consequences of introducing artificial terms; further, the influence of the invention, and of various phases in the evolution, of writing and, subsequently, of printing on cognitive values; the influence of artificial written symbols on the evolution of mathematics, logic and other disciplines. We have here an enormous field for research, one which has, to be sure, often been an object of attention, but one which ought to be addressed finally with methods more exact than heretofore. Precise scientific induction, abundantly supported by a skilful selection of data from the entire scientific material available to us, ought to supplant the erstwhile semi-popular reflections based on a small number of arbitrarily selected and superficially observed facts.

One might name a few more possible questions from this general area of concrete relationships between the results of scientific work and circumstances; the examples given up to now, however, indicate sufficiently what an interesting and momentous task here awaits the theory of knowledge. It is obvious that the point here is hardly to remove the creative element from the evolution of knowledge, or to replace entirely the ideal regularity of norms, which logic and methodology would like to see realized in the process of scientific work, with the real regularity of causal relationships. The empirical function of the 'science of knowledge' comes down to considering, studying and designating the actual aspects of cognitive activity which demonstrate a partial limitation of free and spontaneous creativity—a limitation stemming not from conformity to norms in advance imposed by logic and methodology, but from dependence on real circumstances. No doubt these circumstances will hamper in only very modest measure creative geniuses in moments of supreme inspiration; but there are few geniuses, and even in a genius's life the supreme levels in the scale of creativity are seldom attained. For the most part the development of knowledge takes place through the accumulation of more or less important contributions, and in this enormous mass of average facts, neither exceptionally original ones nor simple undifferentiated repetitions, the theory of knowledge can and ought to discover and precisely determine the component of a constant dependence of results upon circumstances whose existence popular reflection and everyday practice permit us to surmise. No doubt, too, scientific work at a high critical level is partly susceptible to interpretation as being subject to the regularity of normative logic and methodology; yet everyone who has devoted himself to it knows how divergent actual scientific procedure often is from ideal norms, which in the fullness of their meaning appear only in the retrospective critique and final formulation of the obtained results. And if we take the entire history of knowledge under consideration, it will turn out that even after removal of the unpredictable element of uncommon originality, logical and methodological regulation can contribute only in small measure to elucidation of the tremendous majority of cognitive processes; causation will probably prove considerably more fruitful than the assumption that the human mind always follows in cognition the route indicated by logic. Naturally, however, one should always bear in mind the limits of applicability of causation.

Causal treatment of cognitive activity

Finally we pass to what is by all odds the most important type of question in the science of knowledge, the concern of which, as we have indicated generally, is the causal elucidation of the actualization of a designated cognitive operation.

Here we must first of all exactly delimit this area of questions from others. In speaking of elucidating the *actualization* of a cognitive operation, we have in mind not the question of its absolute genesis in the history of human thought, not where this kind of operation arose in the first place, in what previous operations it has its source — for that is a question of the creative evolution of cognitive thought elusive to causal elucidation. Our task is a more modest one: it touches only, so to speak, the time and place of the operation's emergence. More precisely, supposing that a certain operation was in fact to emerge in the creative evolution of cognition, then we should seek to explain what relationship obtains between its emergence and phenomena of one kind or another that occurred contemporaneously, previously or in consequence in the consciousness of the same man and the same community. It is not its logical connection with the past and future of knowledge, not its theoretical function, not its ideal role in learning, but simply the actual historico-biographical fact of its accomplishment that is here the point of departure for our interests.

And furthermore, this fact of its accomplishment can only be grasped as a link in a causal relationship if in principle we can regard it as being typical and repeatable in its type, discovered or at least discoverable in other times and places. For only then will its realization in the given time and place present itself not as a chance consequence of an exceptional coincidence of which human creativity has spontaneously and unpredictably taken advantage, but as a fact connected in a necessary fashion with other ascertained facts. On the other hand, however, since, any cognitive operation interests us as theoreticians of knowledge only when it is relatively original and its next repetitions, unless they are again fairly original modifications of it, possess comparatively little significance for us, as we have already stipulated in the previous section, we are interested not in the one special operation but in the entire class of operations, more or less original on a certain scale of potential originality. Therefore we will attempt to explain not merely a certain individual fact of the accomplishment by a certain individual in a certain society of a certain operation A^1, but any and every fact generally of the accomplishment by anyone, anywhere, of an operation of class A, be it A^1 or A^2 or A^3.

But that is not the end of it. We have mentioned previously that the complete, concrete process of creating a new cognitive value does not lend itself to formulation in a causal law and must be divided up. Thus we pondered first the relationship of the value produced by a cognitive operation to the given circumstances with which the operation has been working, and we took the actual operation indispensable to the production of the value in the given circumstances as a constant component of the process, not considering its possible changes outside the limits of differentiation possible for the given class of operations and not attempting to elucidate the fact of its accomplishment; for it is only on condition of the hypothetical immutability of the operation's class and of its assured accomplishment that the influence of circumstances on the result can be stated exactly. And at present, when the above assumption itself is to be formulated as problematical, when we are concerned with elucidating just this fact of the accomplishment of a certain operation and consequently when we recognize the possibility of its not being accomplished or of an operation of another class being accomplished in its stead, we must eliminate the problem of circumstances influencing the result independently of the nature of the operation. Therefore we consider here both the circumstances and the results of the operation only insofar as they are determined by the operation itself. This means that we prescind completely from the result which the operation actually attains; we are concerned only with the result that it *wants* to attain, with its cognitive intent, in the presumption that this intent has been defined by the operation and that it possesses everything necessary to the realization of that intent. And we take the circumstances into consideration only to the extent that they are already adapted to the attainment of the intent, are already determined cognitively with regard to the result that the operation wishes to attain; consequently we assume that the operation has already selected its object, its materials and tools and that it has posed the question which it intends to solve, so that the concern now is only with solving that question.

We term the cognitive operation thus manifesting itself in its intent, disregarding the question of the agreement of intent with execution, the *cognitive purpose*. Hence the purpose is that element of the operation that characterizes it in its subjective, psychological aspect, apart from the objective role that it carries out with respect to objects given to it with which it works. And we term the system of objective elements determined by the operation, within whose limits and on whose basis the given question has been posed, that special combination of object, materials and tools of which the operation intends to make use in carrying out its assignments, the *cognitive situation.*

Clearly the purpose constitutes the most essential element in the cognitive process; it is preeminently on the purpose that the process depends, hence the purpose is the chief basis for the process's definition. Although the same purpose may express itself in different processes and lead to different results, still these differences are secondary compared to differences in creative processes stemming from dissimilar purposes. For example, the purpose of quantitatively determining the ratios of elements in chemical combinations and the purpose of physico-chemically elucidating, biological phenomena have expressed themselves in different methods and results, but both characterize a certain specific class of cognitive processes. Individual manifestations of each of these purposes differ less among themselves than any of them does from any manifestations of the purpose of qualitatively deducing the 'essence' of a chemical combination from the 'essences' of its components or of the purpose of vitalistic explanation in biology, and the more so from manifestations of any purposes of discovering truths of a certain kind about astronomical or psychological phenomena. Similarly, too, in a cognitive situation we find that which is most essential to the characterization of the circumstances of a given cognitive process, since that is what ultimately conditions the solution of a given question; differences in circumstances originally given to cognitive operations are comparatively secondary so long as an operation of a certain kind is able to produce similar cognitive situations from them, and their importance is the greater, the more diverse are the situations that they suggest to cognitive thought.

The questions that we can put concerning the causes of the emergence of certain cognitive purposes at certain moments in the lives of individuals belonging to certain social groups, must rest on a proper understanding of the nature of psychical causality. Every purpose occurring in the consciousness of a subject (of an individual or a group) requires for its causal elucidation the cooperation of two factors: of another, preexisting purpose in the same subject; and a change in the objective set of phenomena given to the subject that presents itself to him as being pertinent to his preexisting purpose. Neither the psychical process by itself nor the action of 'external influences' by itself can be the cause of a psychical fact. For every psychical process refers to objects of some kind, is intensional, possesses the character of a purpose. It becomes a factor in the emergence of a new process only when there occur, in the objects to which it refers, changes not covered by its intent, hence depriving it of its original conscious role. And external influences are only true influences, they only act on the psychical life of the individual (or the group) if they fall upon prepared ground, if they happen upon some actual or potential intent of the subject's, if the latter gives them some meaning for the sake of his own purposes—in short, if irrespective of the subject's intent they modify some ready *situation* of his. The nature of the imprecisely termed 'reaction' which the given external influence will evoke, depends precisely on the nature of the purpose which corresponds to the situation modified by that influence.

Thus we may say one of two things: either that an emergent new purpose b is a result of the cooperation of a previous purpose a and of an external purpose on a situation A associated with that purpose, or again more simply, that the exchange of purpose a for purpose b (or for a combination $a+b$) is a result of a change in situation A associated with purpose a. It is understood in this connection that a presupposition for this causal relationship is an association between situation A and purpose a, that is to say, membership by purpose a and situation A in the same active process. Thus a new social purpose (e.g., a desire to humiliate an acquaintance) will emerge in an individual when the individual's social environment 'reacts', in a manner not intended by the individual, to some other social purpose, disclosed externally or not, but expressing itself in some social situation of which he is aware (e.g., if the acquaintance reacts by scoffing at an attempt to win his admiration). Independently of the individual's intent, changes that have taken place in the tools or material which the individual intends to use in a certain technical activity, evoke in him a new technical purpose (e.g., a desire to learn a different way of operating with given material) or economic purpose (e.g., a desire to acquire a better tool), etc.

In the same way the cognitive purposes of the individual (or group) are explained causally by the influence of objective changes connected with certain preexisting purposes in the individual. No external factors, natural or cultural, will produce cognitive purposes in the psyche out of nothing; they can only exert a modifying influence on already existing purposes, replace them or complement them with new ones. On the other hand, so long as we stand on a causative ground and ignore the creative element in spiritual development, we have to say that within the limits of causal elucidation no cognitive purpose will emerge in the individual (or group) simply as a result of previous purposes without some objective coefficient, some change in the preexisting cognitive situation. [4] And wherever and whenever a certain cognitive purpose of an individual (or a group) encounters a certain change in the cognitive situation independent of the intent of the individual or group, a necessary result will be the emergence of a different purpose.

Let us now consider some categories of questions that the theory of knowledge can pose and solve with the help of the above premises.

(a) *The influence exerted on knowledge by other fields of human activity.* This question has long been familiar but has acquired a greater significance and broader application in connection with the biological theory of consciousness in general and of cognition in particular, and further in connection with pragmatism. From the standpoint of biological evolution, conscious activity in general has been treated as a broadening and extension of the material or energetic processes in which the adjustment of a living being to its environment manifests itself. This broadening assumes in the species the form of instinct, in the individual — the form of habit. Cognition is a specific function of consciousness that comes into being when, as a result of a change that has come about in the environment, an instinctive or habitual activity encounters obstacles that make difficult or impossible the effective conclusion of its established course; the consequence of such an impediment is reflection seeking to adjust the manner of action to the new circumstances, and a fundamental moment in that reflection is the purpose of identifying the circumstances from the standpoint of the needs of the activity. This biological theory of cognition has been linked in pragmatism with the results of observing more complicated manifestations of cognitive thought accompanying various forms of cultural activity—technical, economic, religious, etc.—which, in contradistinction to theoretical thought, have been given the common term of 'practice' or of practical activity. From the mentioned observations pragmatism has drawn the conclusion that all new, original cognitive thinking is evoked by some practical difficulty requiring solution.

We need not argue how momentous this matter is to the theory of knowledge. If in fact the final and complete causal explanation of every cognitive purpose lay in some other field of human activity, knowledge could be explained in its entirety as being a result of the gradual broadening and accumulation of our entire culture, which in the course of its broadening and accumulation continually encounters difficulties that provoke theoretical reflection. Knowledge would not then be a separate field of culture with its own order but a set of phenomena of the most diverse origin, growing up on the soil of various other fields of activity and organically associated with them.

Precise analysis of facts adduced in support of this thesis, however, readily shows its one-sidedness. New cognitive purposes do indeed emerge, among other things, in connection with questions of 'practice' (in the broad sense of that word as used by pragmatism), but only as changes in previously existing cognitive purposes. For instance, the failure of some practical intent due to altered circumstances compels us to ponder those circumstances, to reflect theoretically, but only when our intent has been linked up to a theory of reality to which that intent has referred, and our activity has been accompanied by the expectation that new experiences will confirm the theory. It is exactly to the extent, and only to the extent, that a practical activity carried out is explicitly or implicitly treated by theoretical thought as an experiment which is to confirm a hypothesis, that is, to the extent that the practical activity is accompanied by a theoretical purpose for which the intent, circumstances and process of practice constitute a *theoretical* situation — that unexpected difficulties encountered by practice strike us as having theoretical significance, as modifying the theoretical situation. It is only through this that they exert an influence on our reflections, that they lead to the emergence of a new cognitive purpose in lieu of, or as complement to, that which appeared in our original hypothesis and was impeded by an unforeseen fact discovered or produced in the course of the practical activity.

The far-reaching significance of practice for the evolution of knowledge does not, therefore, rest in practice being a wellspring of theoretical thought but in its imposing, on the theoretical thought associated with practice, unforeseen changes in theoretical situations, in its not permitting theoretical thought to shut itself up in old problems and solutions but, so to speak, forcing it to notice the inadequacies of its previous theories and to seek new ones. Obviously, however, practical activity only carries out this role if, firstly, it makes use of theoretical reflection in order to attain its aims, and secondly, if it is itself bold, original and creative, if it not only does not steer clear of difficulties raised by altered circumstances on the road to the realization of old goals but moreover seeks out new goals and new difficulties to overcome. On the other hand, theoretical thought hardly need wait for changes that practice may inject into its ready situations. For not only is theoretical thought capable of independent creative development, but, even when it holds to a plotted direction, when its creative spontaneity falters in any field and consequently it becomes dependent on the causal action of external influences, those influences giving rise to new purposes can come not only from other fields of culture but also, and perhaps first and foremost, from other branches of knowledge itself.

Bearing these reservations in mind, let us nonetheless recall several well-known and empirically determined relationships of science to practical questions, in order to illustrate the matter. We know how materially engineering has contributed in recent times to the development of physics and chemistry, and if we were to reach further back into the past — to Archimedes or even to the ancient Egyptians — it would turn out that this influence was no smaller in the first stages of those sciences. The dependence of the biological sciences on medicine is manifest throughout the course of their evolution: the protracted initial period of magico-religious concepts of life phenomena is explained in large measure by the magical character of primitive medical technique, just as in the later gradual emancipation of biology from religious beliefs the progress of positive methods of treatment played an important role. Until recent times questions of politics and morality have been nearly all-powerful in conditioning the evolution of theoretical thought on societal life, and even the very emancipation of modern sociology from this direct dependence on practical life is explained not only by the trend of intellectualism proceeding from philosophy and from the biological sciences, but requires consideration as well of the influence of the needs of social practice, which in the face of the hitherto unprecedented complexity that we find, e.g., in South America, demands ever more thorough and detailed theoretical studies of social phenomena in order to be able to base its reforms on their results. History to this day has not freed itself completely from political and practical-social considerations; philosophical problems, as is generally known, for long periods display the action of religious or of antireligious motives; we know that even in mathematics and astronomy progress in the beginning occurred largely thanks to motives of a technical and magico-religious nature.

In all these examples, which could be multiplied further, the generally formulated course of action on knowledge by one or another field of culture breaks down, upon closer analysis, into a series of particular processes of the type that we have described above, each of which should basically be amenable to interpretation as a manifestation of a more or less general law or synthesis of laws. Thus, for example, engineering, medical and political practice, whenever they plan some relatively complex operation, need a preliminary theoretical orientation in the phenomena constituting the object, the material and the tools of the operation. The technical or social purpose is therefore accompanied, as a special cognitive situation, in a subsidiary capacity by the purpose of a cognitive treatment of the practical situation in question, together with an intended course of action; the cognitive problem that emerges here is that of applying the theory on which a plan for practical activity has come to rest, to future facts which the attempt to realize this plan will produce. This always happens in planned action; it is only in an unplanned operation that cognitive purposes do not appear, except perhaps when the unplanned operation encounters unforeseen obstacles to the realization of its intent and must form a plan of alternative procedure.

So long as the practical plan lies within the same limits of differentiation as a former activity of the given type; so long, e.g., as an engineer wishes to build a machine of the same kind as have formerly been built, a physician to treat his patient for a known illness by known methods, a politician to win triumph for his party in the usual way at the elections, so long there also suffice, in the creation of the plan, cognitive operations of some old type, working with ready-made values. The engineer employs familiar physical formulas in a familiar way, the physician makes his diagnosis in accordance with scientific traditions, the politician informs himself by a common method about the people's position on his party's program and candidates and employs traditional premises concerning the influence of popular slogans, arguments, promises, threats, calumnies aimed at opponents, etc., on the disposition of the people. A theoretical problem always exists, but it is so similar to others already solved that science is not interested in it.

Let us suppose, however, that the practical question at hand is a new one, either because the activity involved sets itself new tasks (e.g., the engineer wishes to build a novel kind of machine, the physician to apply a novel method of treatment, the politician to push through a novel social reform) or because there occur in experience new circumstances making impossible the application of the previous procedure (e.g., the engineer, due to war, cannot obtain some needed material, the physician encounters a hitherto unfamiliar clinical case, the politician cannot agitate for his party on account of repression by the governing party). In that case the framing of a plan is always likewise the posing of a certain, at least relatively vital, problem: whether cognitive values which have up to now repeatedly proven true in relation to facts of a known type brought into play by the corresponding practical activity, will likewise prove true in relation to the new facts that the planned new activity will discover or produce. Thus a cognitive purpose comes into play, to be sure one of a familiar category (since the corresponding cognitive values have already on various occasions been tested), but in any case one rousing a certain scientific interest.

If now the plan proves operative, the cognitive values are considered to be confirmed and no new cognitive purpose arises. If, however, on the contrary the plan proves inappropriate in action, does not lead to the intended results (or if, on a higher level of technical reflection, it turns out in advance that the previous cognitive operations and values do not suffice to build a plan with any chance of success), then that failure, modifying the cognitive situation given to theoretical reflection, will take on significance as a new cognitive problem requiring a new cognitive purpose for its adequate posing and solution. Failure of the engineer's plan produces the purpose of a new interpretation or of a deeper investigation of certain physical phenomena, the physician's failure — the purpose of discovering unknown pathological processes or a new interpretation of known ones, the politician's setback — that of deepening knowledge of certain social phenomena or of changing the premises on which rest methods of social action on the masses. And from there on the new cognitive impulses can continue to develop independently of practical considerations, in association with purely cognitive motives springing from strictly scientific interests.

The concern of the theory of knowledge will therefore be to discover, in processes of this kind involving the emergence of new cognitive purposes from old ones under the influence of modifications produced in cognitive situations by an activity other than a theoretical one, regularities in the form of causal relationships of the kind: wherever and whenever a cognitive purpose of class *A* encounters a change *B* produced by activity of a certain category (technical, social, economic, religious, etc.) in a situation corresponding to it, the result will be the appearance of a new cognitive purpose of class *C*. The future will show in what measure the theory of knowledge will succeed in carrying out this task. In any event, only after the attainment of laws of this type will we be able in fact to state precisely the actual results of the influence of particular fields of culture on knowledge, instead of, as today, contenting ourselves with shallow generalizations; only then will we know what can be ascribed in one or another branch of knowledge to the action of practice, and what to the mutual influence of scientific theories. This is also a matter of first-rate importance with regard to the future of knowledge: whether, when, in what departments and within what limits are we to support the association of learning with practice, or rather seek its isolation? What forms most advantageous to knowledge should this association take, insofar as we regard it as desirable at all in the given domain? How to harmonize the needs of science with the requirements of practice? In the present-day state of the theory of knowledge we cannot give a conscientious answer to these questions.

(b) *The education of individuals in theoretical thinking*. The theoretical education that is the main task of our whole educational system takes in two basically diverse matters: the imparting, to youth, of a certain stock of information, i.e., of certain systems of theoretical values, and the development of their capacity for theoretical thinking, i.e., for carrying out cognitive operations. Here we may pass over the first matter, for it affects only the diffusion in society and the fixing in the individual of a ready stock of knowledge, which exerts no direct influence on knowledge itself; the indirect influence of those processes, we shall consider later. On the other hand, the second matter possesses an extremely important significance for the theory of knowledge, since it is obvious that the entire evolution of knowledge depends directly on the education of scientific workers in theoretical thinking.

In discussing this matter, we will pass over the nonsensical, if (or rather because) today popular, opposition of two tasks of the educational system: one general, consisting in preparing youth for everyday practical activities, and one special, limited to a few almost exceptional individuals, aimed at making savants of those individuals. Naturally, we are interested here in the matter of intellectual education, not of education in general; for no one doubts that with respect to the development of character there can be no other differences in methods of education save those linked to individual and group differentiation of youth on the one hand and to differences in their future professional functions on the other, and no one proposes to make any special exceptions for future scientists and scholars as opposed to all the rest. However, a conviction has taken firm root that in the intellectual sphere extreme differences obtain between methods of successfully educating scholars and methods of successfully preparing people for any other vocation. And since future scientists constitute a very small percentage of the alumni of schools of all kinds, not excluding universities, a view has grown up that the educational system should take account above all, almost exclusively, of the requirements of practical life; often in this connection it is assumed implicitly that an individual possessed of the abilities and taste for scholarly work will get along in any educational system. Interestingly, such sacrificing of the needs of intellectually outstanding individuals to the supposed needs of the generality is often considered to be a symptom of and a prerequisite for the democratization of education.

It is not difficult to show that the intellectual needs of people intending to act in any practical field are in their essence completely identical with the intellectual needs of future scientists and differ from them only in degree. There is only one effective method of educating people intellectually, and that is the one that in its ultimate outcome leads to the development of scholars and scientists. The so-called man of action: the official, the legislator, the civic leader, the teacher, the physician, the technician, the businessman, indeed as the priest, the artist and the poet, needs in addition to training in theoretical thinking still another kind of background, in some field of activity basically different from cognitive reflection — a background which, parenthetically, today's professional schools do not give him to an adequate degree. Insofar, however, as he ought to possess a certain measure of strictly intellectual ability, that ability must be basically the same and be developed in the same way as the ability of an individual specializing in an appropriate field of scholarship, only the degree of its development need not be equally high, since the objective here is not further scientific creativity but merely the application of learning already generated.

The trend discussed here, of adapting methods of intellectual education to the needs of practical people, considered to be different from the needs of scientists, has behind it one weighty motive — namely, a reaction against the overloading of youth with detailed erudition, which in life for the most part turns out to be an unproductive, if not a harmful, ballast. But as we shall presently find, that reaction is equally desirable in the interests of knowledge: overburdening with erudition is just as incompatible, perhaps even more so, with the development of scientific creativity as it is with the ability to act practically. The future savant, like the future practitioner, needs above all a thorough mastery of basic information in a given field of knowledge, information that is indispensable both to any further, detailed research in this field and to any practical applications of it. Only on this common basis, one in fact sufficiently broad to fill a minimal program of higher studies, can there arise a further specialization of knowledge, from the standpoint of practical or theoretical objectives, which should be left for the most part to the independent effort of each individual. However, as we have indicated, the main objective of education meant to produce scholars is training in more or less original intellectual operations. The point is that the practitioner needs this training as much as the scientist does. Although the range of originality required of him is considerably smaller, still he must be capable at least of the minimum of creative thought indispensable to the elaboration of plans for new practical questions which, as we have previously seen, includes at least hypothesizing about the applicability of existing theories to new facts. Without that ability he will always remain either a dilettante feeling his way along or a routinist slavishly keeping to beaten paths. And that ability is but a lower rung of the ability to create new theories.

Theorefore, in considering the question of intellectual education, the theory of knowledge should at once champion the highest requirements in this realm, which *eo ipso* already subsume all the lower requirements, and should reffect upon methods of developing independent scientific researchers and upon the premises on which those methods rest. We postulate that the supreme task of our school system in the educational domain is the formation of scholars and scientists and that all other tasks in this domain should be subordinate to this simply because their realization is likewise completely contained within the realization of that supreme task. From the same considerations we will here take under consideration only the highest stage of schooling, that which directly leads to the formation of scientists and scholars — the univestity stage. For it is here that are manifested most fully and in the greatest detail the general principles of strictly intellectual education, which at lower levels is still linked in diverse ways with moral and physical education and with the development of faculties possessing more of an auxiliary and preparatory character (e.g., perfection of the means of social expression of thoughts through the study of language, the development of the formal faculty of observation and reasoning in general, etc.).

The education of an individual to be a scientist reduces basically to two functions: to giving him a certain technique, that is, the ability to prepare circumstances appropriate to his cognitive intents, and to rousing in him more or less original cognitive purposes. The first matter belongs, strictly speaking, to the category of questions discussed in the previous chapter. Exactly to the extent that we know the relationship between the circumstances and the results of research, we will know how to indicate what kind of materials should be gathered together in certain kinds of questions, how to prepare them for use, what material and cognitive tools to use, what symbols to use, etc. The attainment of certain results in certain circumstances depends, however, as we have seen, on the accomplishment of appropriate cognitive operations; the most perfect technique is utterly fruitless without creative thought. Consequently the main task of university education is to influence the development of cognitive operations. And since it is impossible to impose a complete creative cognitive operation on anyone, the task reduces to arousing in the individual relatively new cognitive purposes whose actual realization, assuming that he possesses the appropriate technical information, depends now only on him.

We can describe this function of intellectual education more exactly in the following way. We are dealing with a person who brings with him certain cognitive purposes but not yet the ones, or in any case not all the ones, that are needed in order that he may become creative scientifically in a given field of knowledge, in order that he may be a more or less original chemist, biologist, sociologist or philosopher. Thus we are gradually to bring him, by acquainting him with the previous achievements of the given science, to the evolutionary stage at which such cognitive purposes will arise in him whose realization under certain circumstances, with the help of certain materials and tools, will lead him to elaborating cognitive values that truly augment the scientific attainments in that field. Leaving aside influences of a social character about which we shall speak later, and which develop in the individual a desire for creative scientific specialization and a drive for self-education complementing and frequently taking the place of the pedagogical activity of a professor, the latter has at his disposal, for the speeding up and guidance of the student's intellectual development, the most important resources in the form of intellectual influences.

Intellectual influences consist simply in application of the general principle of psychological causality, that new purposes arise from preexisting ones as a result of modifications in situations to which those preceding ones refer. In the intellectual sphere a new purpose arises if there are introduced into a cognitive situation corresponding to some old purpose new elements making it impossible for the old purpose to produce the intended cognitive value — in other words if, into a problem whose solution seems to us assured, there enter new cognitive values or new objects of cognition requiring new intellectual efforts and a new posing of the problem. A cognitive purpose emerging due to such an intellectual influence can either completely replace the previous purpose — e.g., as when a theoretical critique demonstrates the total insolubility of the problem posed — or it can merely complement the previous purpose, e.g., when it turns out that solution of the problem posed necessitates posing and solving some other problems. In any case, though, the change is a result of strictly intellectual factors, not of ones stemming indirectly from practical changes as in questions previously considered (Section (a)). An essential feature both of intellectual education and self-education is that they rest on a completely different causality governing the evolution of cognitive purposes, that they are possible solely thanks to the fact that new cognitive purposes arise from previous ones as a result of changes in cognitive situations whose source lies exclusively in a theoretical activity introducing new elements into those situations. In education to situations with which a pupil operates, the new elements are introduced by the theoretical (critical or complementary) activity of the teacher. In self-education, which following the completion of one's higher education remains the sole (if we leave aside the mutual interactions of scholars) means of reflectively, consciously guiding the individual's intellectual development, the person himself introduces new elements into his cognitive situations, since his sphere of theoretical activity encompasses numerous and diverse operations, in part mutually independent, and each of these operations can influence another one, modifying its situation.

It is clear that a truly efficient reflective influencing of intellectual development is possible only given an adequate knowledge of the laws governing changes in cognitive purposes due to intellectual factors. This knowledge is more important in education than in self-education; for in the latter a spontaneous, causally indeterminate, creative development of the individual often plays a relatively greater role than does conscious, planned guidance of one's own evolution based on more or less explicit causal premises; whereas in educating others we rely exclusively on the conviction that certain influences will achieve certain results. As is generally known, it is easier to excel in an intellectual field without any guidance rather than by relying entirely on incompetent guidance. To be sure, numerous centuries of intellectual education have produced a certain approximate, empirical knowledge of some general relationships in this sphere of phenomena; how insufficient that knowledge is, however, is shown even by superficial observation of our university system. Although we personally make no claims to a thorough knowledge in this field, nevertheless some errors in that system reveal such glaring contradictions to the most elementary generalizations, striking the eye of anyone who has studied the psychology of intellectual creativity, that we cannot forgo a few critical remarks.

Today's approach to university education, aside from the imparting of technical instruction, rests mainly on two expedients: on conveying, in the form of lectures, the most important results in a given field of knowledge, i.e., a systematic collection of cognitive values in the field, and on scholarly exercises, i.e., on inducing students to work up, in relative independence, certain typical questions from the given field and occasionally to participate in a subordinate capacity in the professor's scientific works. It is surely not difficult to realize that the first expedient in its present-day form is hardly a practical application of the results of theoretical studies on the evolution of cognitive purposes in students but constitutes a relic from times when due to a lack of books the lecture was the chief, and sometimes the sole, means of passing on the accumulated acquisitions of knowledge to the young generation.

Undeniably an acquaintance with previously produced cognitive values is an indispensable *condition* for subsequent creativity, nevertheless that condition can be satisfied far better through properly selected literature than by listening to a lecture. And it is clear that assimilation of scientific results cannot in itself be a *factor* in the development of scientific creativity. A familiar observation surely states this adequately: those who best assimilate ready scientific values in a given field, are most often themselves scientifically unproductive; perfect erudition not only does not constitute a stimulus to creativity but, on the contrary, if left to itself it hinders the emergence of creative purposes. We can easily understand this by considering its mechanism of action. The emergence of a new cognitive purpose requires a preexisting purpose and a change in the situation corresponding to it, which change the subject perceives as making impossible the present satisfaction of the purpose and thus forcing its supplanting or supplementing with another. But in giving the student a ready system of scientific results we thereby seek to satisfy in advance not only the cognitive purposes he already possesses but also those that may arise within him, we attempt to give him a ready answer both to the questions that he has already put to himself and to those that we are now suggesting to him ourselves. Not only do we not attempt to arouse in him purposes which in the ready-made system of knowledge will not find satisfaction, but often we intentionally impede the spontaneous emergence of such purposes by overloading him with erudition, by making him passively repeat the problems and solutions given him, leaving no time or energy for independent thought. [5]

The only justified role for lectures, thanks to which they would constitute a true complement to information drawn from books and would become a factor in the development of cognitive purposes, would be the reconstruction of the operations that have led to the production of cognitive values, instead of passing on those values in a cut and dried form. Every lecture should give an exact description of the genesis of certain truths in a given field of knowledge, in order thereby to show students, based on examples, what is the essence of creative cognitive thinking. Retracing during the course of university studies the most important of the intellectual processes to which one or another branch of knowledge owes its genesis and its present development, the student would thereby learn really to understand knowledge as a lively and developing domain of the human spirit, rather than regarding it as a collection of inanimate formulas or absolute axioms. In the process he would acquire practice in the methodical and critical posing and solving not of fictional questions *ad usum delphini*, which popular pedagogy enjoys posing, but of substantial questions that human thought has in fact encountered and does encounter on its path.

Such lectures might be of two types. In the one, the student would gain a genetic acquaintance with the whole present state of a given science and finally would have pointed out to him the newest still unsolved problems and fields open to future creativity. In the second, the professor, as a scientist or scholar, in discussing a question on which he himself has worked or is working, would give his students an example of his own intellectual creativity, not in the form of ready results but in an exact description of the process that has led or is leading him to certain problems and solutions. To be sure, we know that many scholars are loath to initiate anyone into the course of their work, that they do not like to speak of their hesitations and errors. As far as work not yet completed is concerned, this reluctance may be justified by the personal temperament of the researcher; however, in relation to research already concluded it is all too often caused by fear of losing prestige, a fear somewhat reminiscent of the stance of the old sorcerers and the present-day prestidigitators. Without a doubt, as we shall see later, a certain esoteric atmosphere about knowledge is often useful in the scientist's relationship to society at large; but where a teacher's task is to develop his pupils, such an atmosphere conflicts with that task.

However, even lectures acquainting the student with creative scientific work, though they lead to a true orientation in the tasks of science, are not yet sufficient to develop spontaneous cognitive purposes. For that the student must be faced with questions which are important and vital to him, and which at the same time lie in the direction of development of the science and require new cognitive operation for their solution. The very *raison d'être* of scientific exercises is to gradually develop in the student not only technical skills but the capacity for creativity. However, a flaw of the present system of seminars and laboratory exercises is the assumption that all students possess the same kinds of predispositions for a given science and that for all there is one and the same route, the same line of development leading to independent thought. On this assumption rests the uniformity of the exercises, the commonness of the topics, be it in the reading and discussion of the same work or in the assigning of the same laboratory experiments or in the working up of the same scientific materials. At best, in the process the student himself is expected to show an ability to select that part of the topic, the work or chapter, that aspect of the materials or questions that especially interest him or of whose elaboration he feels himself more capable; at worst, the same point of departure and the same course of research schematically set by the professor is imposed as compulsory upon all.

Now, it is obvious that despite our absurdly leveling system of secondary education, every student brings with him to the university a different cognitive disposition, and this diversity does not coincide at all with the differentiation of scientific specialties. Although in everyone there may potentially exist a certain minimum of cognitive purposes which he shares with others, the actual significance of particular purposes, of particular interests in his overall intellectual life and the relative degree of his development in particular intellectual directions are usually completely incommensurable in different individualities. Thus the only effective method would be to discover in everyone his dominant interest, his most current purposes, and gradually to develop, differentiate and complement those purposes in order that, proceeding from them, he may gradually take in the most important departments of the field of knowledge, thus binding them into an organic whole under the action of his most vital intellectual current; arriving at the stage when every new cognitive purpose that may emerge within him will manifest itself in objectively original creative work. And with that intent he should be given problems which, while lying in the direction of his purposes, nevertheless compel him to constantly expand the sphere of his previous interests and knowledge and demand an ever more independent performance of ever more operations, until at last he attains that level at which he will no longer find models in the past and will have to continue on his own.

In the intellectual development of the individual, however, there are acting not only cognitive factors that are either spontaneous or associated with practical questions. An individual who plans a practical activity or seeks to master and develop a certain field of knowledge is not the isolated subject of an operation or thought but a concrete member of social groups which impose on him certain requirements involving every aspect of his conscious life, and hence also of his intellectual activity. Thus, for example, the directions and methods of intellectual education — such as those which today predominate in our school system and of which we have here given a critique — are in every age and in every community the result of certain social traditions and purposes, and today's university student and professor, much like the pupil and director of a philosophical school in Greece, much like the ancient Egyptian priest and his disciple, cannot be understood as social individuals otherwise than against the background of their social environment. That brings us to a new extensive field in the theory of knowledge.

(c) *The question of the social determination of a scientifically active individual.* Cognitive activity, fairly early in the history of culture becomes a *social institution,* that is, certain individuals specialize in it temporarily or permanently, and this specialization is subject to sanction by the group to which the individuals belong; the cognitive activity of the individuals is treated by the group as if it were carried out in the group's name, by its consent and for its benefit, much as any professional work or special function. Here above all we must distinguish the permanent profession — in which the individual devotes his whole life partly or completely to knowledge, as does the shaman (priest, sorcerer and physician in one person) in lower tribes, later the priest, then the philosopher in the Greek sense, and finally the savant-specialist — from the temporary specialization, in which the cultivation of knowledge is linked to a certain age class, as during preparation for other social functions (the pupil), or on the basis of experience acquired in the execution of other social functions (the old man as social adviser). A further important difference occurs between involvement in esoteric knowledge — i.e., that limited exclusively to circles of professional specialists — and involvement in exoteric knowledge, intended for dissemination among society generally; and finally, both with esoteric and with exoteric knowledge the social function of cognitive activity may consist chiefly in the preservation and transmittal of existing knowledge (as in ancient Egypt and medieval Europe) or in the creation of new knowledge (as in Greece and modern Europe).

The social determination of individual cognitive activity manifests itself preeminently in the social group — by its positive opinion, by the glamour investing the position of the professional theoretician, by a guaranteed living for the individual who devotes himself full-time to knowledge or by attachment to other social positions that secure a living, a precondition for acquiring a certain knowledge, by diplomas and special awards and honorary or material distinctions.— associating certain values generally regarded as being positive with manifestations of cognitive purposes which it treats as being desirable. Conversely, condemnation by group opinion, deprivation of professional position, the creation of financial difficulties, and occasionally the imposition of a material penalty (from detention of a lazy pupil after school, to the death penalty for a dangerous thinker) are methods associating values generally regarded as being negative, as the case may be, with lack of signs of cognitive purposes in an individual of whom the group requires those signs, or with signs of cognitive purposes which the group happens to condemn for any reason. These methods cannot, of course, directly produce or remove the cognitive purposes that they have in view, since cognitive purposes, as we know, change only under the influence of cognitive situations, and neither positive nor negative social pressure produces cognitive situations. Nevertheless, they do have an indirect influence, and one of two kinds.

Acting on other purposes of the individual's — social, financial, hedonistic — they cause the individual, in order that he may attain positive values in these realms or avoid negative ones, to make use of certain cognitive values or to shun certain cognitive values: in order to pass an examination he absorbs certain scientific or scholarly results, in order to obtain a scientific or scholarly position he displays his erudition in the required field, in order to win popularity he occupies himself with popular questions and does not touch truths socially dangerous for him, etc. In this way, to be sure, the character of his purposes is not (but the sphere of his cognitive values is) shaped to conform with the requirements of the milieu, being the more explicitly delimited the better are the requirements defined, as to content and significance, of the intellectual values desired by the group. And if an individual of this kind occupies himself with knowledge professionally, operating constantly with a certain domain of values, sooner or later he undergoes an evolution in this field, and some more or less new cognitive purposes will arise within him. Under the influence of these same motives, giving the broadest possible expression to purposes of his own whose manifestations the milieu sanctions positively, and suppressing the active manifestation of purposes negatively assessed in his society, such an individual can become a fairly productive professional, exactly adapted to the cognitive needs that his group currently feels.

Understandably, though, the range of his creativity will not be great. For the emergence in him of a new purpose depends on whether, in the domain of values imposed on him by society, there occurs a new cognitive situation that will in fact stir some preexisting purpose of his and effect a change in it. Whereas it is well known that commonly in sets of cognitive values imposed on the individual by society there are no frankly and conspicuously new situations, that they are most often ready-made and solved problems and that only either unexpected perturbations in a group's traditional cognitive scheme unwanted by the group (e.g., the demonstrated uselessness of acknowledged truths in the planning of new practical questions, or the influence of values brought in from other societies) or the creative initiative of individuals independent of tradition can introduce genuinely new problems into the domain of socially acknowledged knowledge. And even if a new purpose does emerge in such an individual, it can never become revolutionary, it can never develop to such a degree as to transform fundamentally some basic part of the traditional sphere of cognitive values, since the social group always fears revolutionary transformations in any field. There are, to be sure, periods and societies in which, nominally, originality and even revolutionary views are positively appraised by the community; closer study, however, shows that only those scientific revolutions (as, indeed, sociopolitical revolutions) enjoy immediate acceptance by society at large, in which there appears not a fundamentally new direction of thought but either merely a new formulation of familiar truths or a return to some earlier direction, apparently forgotten but in reality still strong among the masses. All of us are familiar with the type of person that might be termed the 'cognitive philistine', openly conservative in societies that set obstacles to any and all innovations, superficially original in groups which in the intellectual realm are subject to fashion rather than tradition. This type flourishes mainly on the soil of exoteric knowledge, that which is imparted in the schools to youth (not specializing in scientific or scholarly work) and propagated in popular lectures and popular science writing, etc.; for only exoteric knowledge can be subject to the kind of social control of which we are speaking here.

The influence of social circumstances on the individual's cognitive creativity can, however, be different, deeper and more fruitful, if again only indirect. Namely, the group does not always nor over the full range of knowledge set specific requirements as to the nature of the cognitive purposes desirable and permissible in the individual. Its demands are very clear in regard, e.g., to the range and kind of knowledge that is to be assimilated by the generality of youth, and generally enter imperatively into the intellectual life of those members of the group whose cognitive operations do not constitute their total and exclusive task in life. On the other hand, they usually become less strict, more general, in respect to individuals whose cognition is supposed by the generality to rise above the level of popular comprehension. The cognitive professional, the priest, the philosopher, the scientist, insofar as he has not occupied himself with popularization, has always been surrounded by a certain aura of sanctity, mystery, inaccessibility; every society has not only permitted but in some measure even viewed positively a certain esoterism in cognition. This esoterism, at first based chiefly on an external sequestration of the cognitive elect from the crowd, and later stemming simply from the internal inaccessibility of the deeper cognitive values and operations for the unprepared, has always left, at least to some individuals, a relative freedom of scientific or scholarly action within more or less broad limits.

This freedom may again sometimes have been hampered by esoteric associations, by special groups formed within the broader society for the cultivation of esoteric knowledge; we shall have more to say of this later. In any case, though, the associations themselves have usually defended their members from excessively strict demands and sanctions by society at large, have guarded the independence of knowledge from popular opinion; while on the other hand the individual has sometimes been able to appeal to the broader society against pressure by the association, when that pressure has been excessively tyrannical, when the association has been too ossified in tradition whereas broader social spheres had undergone a certain, if inadvertent, evolution. In this, as in any, field of cultural life, simultaneous membership by the individual in two or more groups has been conducive to the individual's liberation from overly exclusive dependence on one group, has enabled him to attain a certain sphere of private personal freedom. This sphere has further grown markedly in more recent times thanks, on the one hand, to the plurality and differentiation of national societies possessed of a more or less intensive scientific life, and on the other to the contact in which the scientific associations of the various countries find themselves.

This freedom in the realm of cognition has never been a license free of all regulation. It is true that often individuals and even small groups have taken advantage of some cognitive esoterism, or even artificially manufactured such an esoterism, for the sake of scientific charlatanism; still, the 'charlatan' in the realm of learning, like the hypocrite in the religious and moral realms, is only an aberration from the norm that in itself points to the existence of the norm. By the very fact that professional cognitive activity was institutionalized, that society had, as it were, officially recognized the priest, the philosopher, the scientist as an individual in a way exalted above others, this recognition was accompanied by certain demands of a general nature which such an individual had to satisfy in order to preserve his right to esoterism. The broader social group, nation or state, as well as the esoteric association, have placed before the cognitive elect a certain personal *ideal* whose realization in life has been the price of his exaltation above particular observances and constraints to which exoteric knowledge has been subject; the generality's conviction that the cognitive professional submits to that ideal, has been the basis for their confidence in him, even when they have not understood his activity.

The personal ideal of the 'thinker' (if we may use this term in common for all the afore-mentioned classes of cognitive specialists) has been different in various societies and ages. Often, also, an ideal prevalent in an esoteric association has been somewhat different from the ideal acknowledged by the society that has sustained the association, and an individual who, in his struggle with the generality, has leaned on the association, or who, in his struggle with the association, has appealed to the generality, has championed one of these ideals against the other. Nevertheless, despite all these differences, the community of the traits is marked enough for us to be able to consider these temporally and locally differentiated ideals as variants of one basic ideal. Negatively it is always defined by the requirement that the thinker be independent of those values that are the chief motives for the activity of the generality — first and foremost the economic and hedonistic ones, and in some measure even the social ones which manifest themselves in the search for power or popularity. It is characteristic that this is exactly the criterion differentiating the esoteric thinker, whom the generality is prepared to trust and vouchsafe a certain freedom, from the exoteric-cognitive philistine, whom it does not trust and whom it controls at every step, although on the other hand the philistine is often a favourite of the masses and recipient of the most rewards, whereas the thinker tends to arouse in the masses a cool respect and a desire to isolate.

Positively the ideal of the thinker unites within it two requirements: excellence in carrying out his special function, and public service, or action for the good of the group. The criterion of excellence, obviously, will be different when the main task of knowledge is the preservation of old truths, and different again when the objective is the discovery or creation of new truths; usually, however, the two elements are conjoined in various proportions. In any event, the actual criterion of excellence is not imposed by the generality, but is developed by the thinkers themselves and fixed by the professional associations. The requirement of public service was satisfied to the highest degree, in the conviction of the generality, by priestly knowledge, inasmuch as it was on the latter that the capacity to obtain the beneficent influences of religious and magical powers for the group was deemed to rest. Secular knowledge became public service when the generality began to realize that it was such knowledge that was the source of the intellectual power indispensable to any improvement in the sphere of material and social practice. Hence the demand, universal today, that esoteric knowledge sooner or later be applicable in practice, and the growth, alongside the class of pure theoreticians, of a class of technicians mediating between knowledge, incomprehensible to the generality, and the needs of the collective life.

These two requirements — of excellence in the domain of cognitive values and operations, and of the practical applicability of cognitive results for the good of society — not always mutually harmonious, and mutable in their content, nonetheless constitute guidelines for the individual's development as a thinker, and that is their significance for knowledge. The purpose of realizing the social ideal of the thinker in one's own personality arises as a social purpose, as a complex result of numerous influences by the social milieu to which the individual educating himself to be a thinker is subject from the first moment when there begins to take shape in his consciousness the intention of specializing in a cognitive activity. However, that social purpose gradually leads the individual outside the realm of social life in the strict sense of the word, into the domain of objective theoretical criteria. Desiring to conform to this ideal from social considerations, he develops a conscious measure of his own intellectual progress, and with time that progress in itself becomes his ultimate task. And despite the fact that, as we have said in the previous section, techniques of educating scientists and scholars and techniques of self-education are still of a comparatively low standard, yet in view of the fact that the existing standard of knowledge is also, with few exceptions, still low, the individual, given intensive and persistent pursuit of the ideal of cognitive excellence, by way of numerous wasted efforts and vain attempts, nearly always reaches the standard at which his further activity is necessarily the creative addition of something to the existing store of knowledge. And then the social influences that were the first cause of his purposes may to such a degree be neutralized by the influence of exclusively intellectual factors that the individual sometimes consciously opposes both the broader society and the professional association, in the name of the very ideal of cognitive excellence that those groups had initially fostered in him.

As we see, here again there is rich ground for the theoretician of knowledge. Research into the regulation of exoteric knowledge and into ideals of esoteric knowledge, in connection with comparison of the various social types of cognitive philistines and of thinkers that we find in history, will permit precise statement of the actual influence of social conditions on the scientific worker, determination of the indirect consequences of this influence for particular fields and varieties of cognition and the drawing therefrom of practical conclusions for the future. Of course, here too, as everywhere else, exceptionally creative and original individuals resist total confinement within any framework of theoretical generalizations and practical guidelines; in fact, we find creative scholars and scientists outside the circle of professional and socially recognized workers. But every individual owes something, positively or negatively, to social influences; in some measure his evolution is explained by the action of the social environment, if only because he has rebelled against that environment. Hence it is only a matter of degree, how far elucidation through social influences permits understanding his spiritual process.

(d) *The question of the intellectual life of social groups.* The relationship of the individual's intellectual life to his social environment, considered above, leads us on to further questions deriving from the borderline between the theory of knowledge and sociology. The concern now is the creation and the adoption by the social group of those common cognitive purposes which find expression in the collective recognition of a set of values and in the imposition on individual members of certain rules of thinking and a course of intellectual evolution.

Here again we have to assume the existence, in members of the group, of cognitive purposes preceding the development of a collective intellectual life and of any cognitive evolution generally of the group as a whole. These may be purposes innate to members of a genetic group, i.e., of one based on common origin and taking in, from birth, all descendants of individuals belonging to the group at the moment (as, e.g., a family, tribe, nation); they may be purposes partly innate, partly acquired in members of groups formed via co-opting older individuals (e.g., a school, a scientific society). It is understood that these cognitive purposes preceding the collective intellectual life of the group are never uniform in all members of the group; for not only may there exist, in various individuals, purposes qualitatively more or less at variance with those that other members possess, but even the purposes that are found in all, may manifest themselves differently among them. In some they are more lasting, actualize more frequently and consequently play a greater role in the totality of their conscious life, whereas in others they are transitory, seldom actualize and consequently have a smaller personal importance. Later we shall put the question as to whether and to what extent we may nevertheless assume, despite all these individual differences, a relative uniformity of primitive cognitive dispositions in certain groups. Of course, in any case such a primitive uniformity can only be partial; no innate similarity among members of the group, by itself, would ever suffice to explain the community of the convictions and cognitive purposes that sometimes reign omnipotent in the group.

The community is hardly explained by the fact that all the members of the given group are subject to more or less like external influences, be it from the natural environment or by the action of other groups. For like influences produce like results only insofar as they are linked to like purposes; given a diversity of individual purposes, the results of like actions will of necessity be diverse. If we transfer various individuals separately into the same natural or social environment; then their differences, instead of blurring, will become still more conspicuous. Only if such diverse individuals form a unified social group will common new circumstances produce, in addition to divergent results in every person, also certain uniform results in all. This means that the sources of social uniformity lie in something that is to be found only in the social association of many individuals into a single group. Especially in the cognitive realm sheer identity of external actions obviously does not suffice to produce social identity. It seems even a less improbable assumption that individuals of various dispositions, finding themselves in like natural circumstances, will be led to the discovery of like technical tools, than that there will develop in them purposes of elucidating like phenomena in a like way.

There is a well-known theory, originated by Tarde, according to which uniformity of 'beliefs', similarly to uniformity of 'desires', to the extent that such uniformity is not innate, constitutes a result of imitation. This theory, however, treats social processes too superficially. Above all, why do just these beliefs spread, and not others? If there is no special reason inclining all the members of the group, regardless of the individual differences in their purposes, to accept a certain belief while rejecting others, in that case the choice of a belief imitated by each individual will be conditioned on the one hand by objective factors (by the nature of the belief or by the personality of him who initiated the belief) and on the other hand by the subjective inclinations of the individual: various individuals will imitate various beliefs, and no exclusive uniformity of certain beliefs will ever come about. And, even when a certain number of individuals choose the same belief, the latter will take a different shape in each of them and, if there is no reason to consider all these variants as subjective interpretations of the same objective belief, we shall once again have diversity instead of uniformity. In short, imitation would lead to uniform results only if the imitating individuals were already uniform; the theory of imitation thus assumes implicitly that which it intends to explain.

However, if we pass from this theory to its antithesis, represented most strikingly in Durkheim's school, according to which it is the social objectivity of collective 'presentations' imposed on every individual that is the reason for their acceptance by all individuals, we notice that we again stand before an unresolved problem—how and why just this 'presentation' and not another has taken on that objectivity. Certainly this theory considers that the sole positive, scientific method of solving the problem is to seek the source of the objectivity of a collective presentation in previous collective presentations that have possessed that objectivity. That, however, again confronts us with the problem of the cause of the replacement or complementation of those previous collective presentations by the present ones, in which ultimately there will again come into play the question of the uniformity of individual reactions to particular influences. Furthermore, at every step we find the emergence of new 'collective presentations' not at all amenable to reduction to previous ones, since their emergence is linked with the organization of new social groups that are only in the process of evolving common convictions.

This seemingly insoluble problem, of how collective uniformity arises from individual diversity, is, however, solved with ease if we realize (as, incidentally, Durkheim also emphasizes) that the essential, fundamental thing in the collective purpose of a social group is not that it is uniform in all members, but that it is *common* to them all, and if we inquire about the sources of that community. Community may be associated with uniformity, e.g., as when all the members of a given group have the same cognitive purposes in connection with a religion or with a political life, but it can also coexist with considerable differentiation, as, e.g., in a scientific society which is jointly interested in the development of the same discipline, though each member is actively occupied with a different special branch of it while conveying his results to others and in turn having their results made available to him. The only thing that is uniform to all the members here is recognition of the development of the discipline as a common task for all members of the group, whereas modes of carrying out that task are different with each.

Among cognitive purposes present in members of a given community, besides those which each member shows on his own in relation to situations that interest himself above all, we find as well those which refer to situations indirectly or directly *touching the group as a whole*, and it is these that are common *par excellence*. For example, every member of a semi-savage tribe is interested in the magical explanation of events befalling him personally; beyond that, however, both he and all the other members are also interested in the magical causes of events that befall the entire group and are momentous for all. Likewise, at a higher level of development, priestly knowledge is regarded as being a collective value of the entire society, and the cognitive purposes of the priests are regarded as representing the purposes of the generality, since the latter's existence depends on a proper understanding by the priests of the nature of religious powers and of their action. When at last knowledge becomes a secular matter, once again a group — the community, city, state, nation — treats as common, as 'public', those cognitive purposes which present themselves to its members as useful for the whole group. Furthermore, numerous cognitive purposes referring directly to the private situations of particular members or to objects not pertaining directly to the group's existence, assume a public significance through the association that they acquire in the general consciousness with truths pertaining to general matters. For instance, a group takes an interest in theories of economic phenomena if these theories, in the group opinion, can exert an influence on certain social institutions, and recognizes physical and biological hypotheses as a common matter if they are linked in its eyes with religious dogmas which have long since had significance as common values.

Thus the source of the community of cognitive purposes in a social group, and consequently of any similarity in convictions and modes of thinking that comes into being despite the diversity of cognitive dispositions introduced into the collective life by each member, is the community of societal interests of the individuals comprising the group. Recognition of certain cognitive purposes as being common is bound up with recognition of the situations corresponding to those purposes as being important to the group as a whole; and that does not in the least require *uniformity* of individual preexisting purposes, but their *synthesis* for the collective existence. Naturally, as regards those situations the attainment of unanimity is essential in the eyes of the group; that uniformity, however, need not be an actual uniformity of cognitive purposes, only a uniformity of *convictions*, i.e., of a general recognition of certain purposes as being positive in connection with certain situations.

Decisive to the collective conviction is the position taken in regard to a common question by individuals who are, in a sense, authorized to think cognitively in behalf of the group, whose social function consists in posing and solving such questions. Any other member is demanded merely to accept that position as the position of the group, much as the military decision of the commander must be recognized as binding upon all by each of the warriors subordinate to him—which obviously does not mean at all that they must always imitate his acts. 'Imitation' in the cognitive realm, insofar as it occurs at all, springs from a conviction based on recognition of the opinion of the intellectual leader. An individual who has accepted as true the professional leader's manner of posing and resolving a question involving the group — to the extent that he is going to think about that question at all — is going to be subject to the influence of the collective conviction and, in posing and solving his problem, is going to strive for a result compatible with the recognized one. However, this will not be imitation in the sense of repeating another's operation, but only in the sense of imitating its results. Hence we may speak of a common cognitive purpose expressing itself in the treatment of a common situation, although actually that common purpose may be the result of a synthesis of diverse individual purposes striving to conform to the purpose of the intellectual leader.

In the same way that a common purpose manifesting itself in the posing and solution of a common problem binding upon all, has once been formed and fixed, it can later change. Some individual may pose and solve the problem differently or replace it altogether with another; his new rendering of the problem either will be rejected or will again be accepted as binding. Both alternatives depend not on whether this rendering of the problem encounters uniform cognitive purposes in all, but on whether in the overall conscious life of the group a more important role is actually played by cognitive or social purposes which find their satisfaction in the older rendering of the problem or by purposes that, for one reason or another, are satisfied by the new rendering. The sign of victory for the second set of purposes will be the imposition, upon all members of the group, of an appropriate new conviction under whose influence there will again take form, from the diverse purposes of the individuals, a common collective purpose. This schematic representation of the matter should not, however, suggest the conclusion that every collective purpose is originally the creation of some one individual. Most often we find gradual, imperceptible modifications and complementations of existing common purposes by small individual transformations and additions, so that it is impossible to distinguish the part of particular intellectual leaders in the posing and solution of a collective question; neither they themselves nor the group are sometimes aware of the changes taking place in the collective intellectual life.

The matter of the emergence and evolution of collective cognitive purposes assumes a different character when the concern is not with common recognition of some particular question and solution, but with the common conscious creation of new truths, as in associations organized for the collective development of a discipline or group of disciplines: in academies, scientific and scholarly societies, partly in universities, etc. Although here basically every member is required to create new truths on his own, and hence to develop new cognitive purposes for himself, nevertheless in these associations too we usually find, if not community of problems and solutions, then at least a unanimity as to the direction of evolution of the discipline and as to the meaning of the results obtained. To be sure, it is often implicit in the founding of the association that the proposition that the only common cause of all its members is the very development of the discipline and not the conferring upon that development of a specific framework or results, and that every new cognitive purpose of any member working in the common field is to win recognition from all. In reality, however, the results of cognitive purposes of particular members may clash, and in resultant conflicts the group adopts a position. In this way, through a series of such acts of collective recognition, there comes into being a certain common line of development and there emerges a framework, at times a very rigid one, beyond which the cognitive activity of the association does not extend.

Up to now we have considered a social group in abstraction from other groups, as a separate, self-contained entity. However, the fact that the intellectual life of a stated group, whether it be a nation or a scientific society, stands in an objective relationship to the intellectual life of other nations or societies, and at the same time differs from them, holds far-reaching significance both for the theory of knowledge and for sociology. In this relationship and opposition to other groups, the sphere of intellectual culture of a given group becomes as it were the collective property of the group, something that appertains to the group and sets it apart; and consequently the group strives to bind that sphere to the domain of its purely social interests, introducing into its intellectual life motives of internal solidarity, collective egoism, antagonism to other groups, collective pride, group idealism, etc.

The commonest manifestation of this association of cognitive with social purposes is the practical emphasis that the group places on those features of its intellectual life that distinguish it from other groups. Every group is prone to an exclusively positive appraisal of its own culture and a negative appraisal of an alien culture, solely because the latter is alien and different from its own. This is linked with a desire to isolate itself from alien influences, to preserve its own individuality. In the cognitive domain these desires are perhaps less conspicuous than, e.g., in the domain of religion or manners and morals since, in the intellectual sphere, a critical attitude and objective tests of values are more easily developed. Nevertheless here too, at every step, we encounter attempts to emphasize and enhance the group's individuality, either on the basis of an actual sense of the superiority of its own concerns, its own modes of thinking, its own results over those of others, or at least by artificially fostering in the group's members a faith in their own superiority. We need not reach to the remote examples of China or of ancient Egypt for evidence: each of the modern civilized nations, every learned organization, will supply numerous facts of this type. This subordination of the intellectual life to social considerations reaches its apex when a nation's knowledge is used as a tool in a struggle against other nations and when scientific truths are consciously twisted to the purposes of social propaganda.

Narrow exclusivity and antagonism with respect to other groups is transformed into rivalry in the scientific field when the intellectual spheres of various nations or various cognitive associations approach one another in a fundamental recognition of all knowledge as being the common property of *all civilized mankind* and when the sense of distinctness of the nation or association, instead of manifesting itself in a recognition of its own cognitive purposes and values as being absolutely positive as opposed to those of others being negative, expresses itself in a desire for one's own nation, for one's own association to add more than do others to the common treasury of knowledge. That to this end it is oftentimes desirable to cultivate certain special interests, methods and procedures of research, a certain collective specialization, is beyond doubt; acknowledgment of that fact, however, need hardly be linked to a hostile self-isolation from other groups but, on the contrary, can bring in its train an intensification of efforts to develop the best possible acquaintance with, and as objective as possible an appreciation of, the intellectual life of others.

(e) *The question of innate cognitive dispositions.* In the deliberations of the previous sections we have sought to outline a general framework for study of the influence of practice on theory, of theoretical education, of the social determination of individual cognitive types, and of the social conditioning of the collective properties of groups in the realm of cognitive life. There still remains, however, the most difficult and just possibly the most momentous question: *what, if any, innate cognitive dispositions should be posited in individuals and groups*? These dispositions being independent as well of the influence of practical questions, as of methods of intellectual education, and of social circumstances, but constituting, as it were, the basis for action of all cultural agencies. There are chiefly three problems involved here: whether mankind possesses any stock, common to everyone, of cognitive dispositions, identical in all ages and in all varieties of the human species; whether in particular races and nationalities we find hereditary and scientifically determinable cognitive faculties and dispositions distinguishing each race and nationality from others; and finally whether and to what extent the individual differences that we find in the faculties and dispositions of various people can be divided into acquired and innate, and whether on the basis of the latter it is possible to divide individuals into cognitive classes or types paralleling or cutting across the division into races and nationalities.

The chief difficulties in this domain derive from the fact that inborn cognitive dispositions, indeed like most other inborn dispositions brought into the world by conscious beings, do not manifest themselves all at once but show up fully only by degrees, in the course of active contact with appropriate fields of experience. And since every human being, from the earliest times accessible to us, has always been subject from birth to certain cultural influences modifying his inborn purposes, at first glance it is impossible to distinguish what is inborn in him and what is developed by culture. Likewise where a group is concerned. The intellectual life of any social group, like any field of its activity, is certainly shaped in each generation on the basis, to be sure, of inherited dispositions, but under the predominating influence of all the intellectual values accumulated by preceding generations, within a framework of traditionally transmitted logical forms and scholarly or scientific methods; nor ought we to pass over the influence of other groups, continually impinging from outside. How do we separate, in the overall outcome, that which is passed down from generation to generation by way of biological heredity, from that which is a result of cultural heredity, of the continuity of traditions, and lastly from that which has been brought in from outside? Nor are we in a better position when we wish to define the cognitive faculties and dispositions of mankind in general: indeed, as far back into the past as we care to reach, we always find a stock of culture common to all peoples, if more or less differentiated, and we cannot ignore the fact that in every period the intellectual life of mankind accessible to us is a result of the interaction of that stock of culture with the biological predispositions of the species.

The numerous essays at a scientific treatment of these questions, whether as deduction of the human species' cognitive functions from biological instincts, or as characterization of the intellectual inclinations of particular races and peoples, or as the well-known classification of individuals by intellectual types, or lastly in still other forms which we need not detail here, largely ignore or fail to appreciate the difficulties mentioned. Therefore their results up to now should be considered as indications of certain general possibilities which future methodical research in time may state precisely and confirm or discard, rather than as actual solutions or even as exact statements of the problems. It is clear that we should begin not from the definition of innate cognitive purposes but, on the contrary, from the study of *acquired* cognitive purposes, which, precisely because they are acquired, are easier to designate exactly and the genesis of which can be empirically traced. Only through the gradual exclusion of that which turns out to be acquired will we arrive at the primal sources of the intellectual life of the individual, the nation, the race or of mankind.

This process of exclusion should make use of two mutually complementary methods: the genetic and the comparative. Studying genetically the causes of the emergence of each cognitive purpose from previous ones under the influence of changes in cognitive situations, we may at length arrive at purposes which will have to be recognized as existing primally, as not being deducible causally from others; in relation to these, external influences would not be coefficients of their emergence but only causes of their passage from the potential state in which they persist prior to contact with experience, into an actual state. However, inasmuch as the innate dispositions of an individual or group may be expressed not only in the existence of genetically primal purposes but also in certin characteristic variants and combinations to which genetically later purposes are subject in the individual or group, then comparison of various purposes in the same individual or group with the purposes of other individuals or groups should work together with the genetic method in determining what we might term the *cognitive character* of the man, the nation, the race or the species.

*

The guidelines for the theory of knowledge sketched above cannot be, of course, either exhaustive or final. Innumerable new, today not even foreseeable, questions will no doubt suggest themselves even as research progresses, and the methodological canon that today, at the inception of the discipline's evolution, seems indispensable and obvious, is also bound in the course of events to undergo unforeseen refinements and transformations, or even to be completely discarded. For the present, the main objective is to frame at least a theory of knowledge as a single positive science based on experimental material and directed toward the most exact possible description, classification and causal elucidation of cognitive phenomena as cultural phenomena given to us on a par with linguistic, economic, social and religious phenomena. Another objective is for those interested in human knowledge as an object for cognitive reflection to realize that this subject may not only be approached from the standpoint of epistemology, logic or concrete and individualizing history, but that there exists another, as yet unexploited vantage point from which there open up to the inquiring mind new, enormous vistas [6].

NOTES

[1] I. Kant, *Critique of pure reason,* 'Transcendental Methodology', IV, translated by J. M. D. Meiklejohn, London–Toronto, 1934.

[2] It is only within the closed confines of elaborated scientific systems that certain operations positively require others.

[3] In the cited sections of Baldwin's *Mental Development* and Hobhouse's *Mind in Evolution* we find a partial application of this method, but as yet without adequate factual basis.

[4] We emphasize once again that this is not a postulate that might in fact be extended to all of conscious life; since free and creative development is a fact, a causeless actualization of new purposes cannot be philosophically ruled out. But the bounds of science end where causal explanation ceases to be sufficient.

[5] I know of nothing more deadly in this respect than the attempts currently prevalent in Poland at a detailed regulation of the entire university curriculum, especially in law and medical schools and in courses preparatory for the teachers' examinations. No future lawyer, medic or teacher who wants to conscientiously satisfy the requirements set him in regard to erudition, and does not want to extend the period of his studies by 2–3 years, has time or energy to think anything through independently during his entire stay at the university. What will come of a generation whose minds in their best age, just when intellectual spontaneity should develop the most, are systematically and relentlessly pressed into a framework of ready-made formulas? Of course, there are individuals whose sense of conscientiousness in preparing for examinations yields pride of place to instinctually felt needs to think independently, as there are others too — and they are the majority — in whom the same sense of conscientiousness is overshadowed by indolence or by a desire for amusement. Still, what is one to think of a system which makes creative intellectual development possible for some only at the cost of unconscientiousness, and in others cannot develop that minimum of intellectual interests which is an indispensable feature of an intellectually mature man?

[6] In addition to the works mentioned in the text we shall also cite, among works that shed light on various questions touched upon here: Devey J., *How We Think*, 1910, and his *Democracy and Education*, New York 1916; Durkheim E., *Les règles de la méthode sociologique*, Paris, 1895, and 'Representations individuelles et collectives', *Revue de la Métaphysique*, 1898; Jaspers K., *Psychologie der Weltanschauungen*, Berlin 1922; Lévy-Bruhl L., *La mentalité primitive*, 1922; Spencer E., *The Principles of Sociology*, 1876–96, Pol. ed. *Zasady socjologii*, 1889–98; Stern W., *Werden und Wesen der Persönlichkeit*, Leipzig 1913; Tarde G., *Les lois de l'imitation*, 1898; Thorndike E. L., *Individuality*, Boston, 1911, and *Educational Psychology*, New York 1913; Tönnies F., *Kritik der öffentlichen Meinung*, Berlin 1922; Znaniecki, *Cultural Reality*, Chicago, 1919, and *Wstęp do socjologii* (Introduction to Sociology), Poznań, 1922. See also for the psychological aspect of cognitive activity, the bibliography in Stern W., *Differentielle Psychologie*, Leipzig 1911, pp. 425–438; for the social aspect, the bibliography in Park and Burgess, *Introduction to the Science of Sociology*, Chicago 1921; and Polish bibliographies in *Ruch Filozoficzny*, *Poradnik dla samouków* (The Autodidact's Adviser), *Dzieje myśli* (*History of Thought*) and *Nauka Polska*.

Maria Ossowska and Stanisław Ossowski

THE SCIENCE OF SCIENCE*

Two points of view in the investigations which take science as their subject

For a number of years science itself has bceome the subject of scientific investigation, both science in its functional sense (the whole of research activities), and in its static sense (the whole of the products of these activities, i.e., the whole of the scientific truths). [1] How vivid the interest taken in this research is, is eloquently proved by the bibliography published in Vols. XIII and XIX of *Nauka Polska.*

In the first, as well as in the second sense, science may become a subject of research in different ways. It seems, however, that two chief points of view may be established to which all investigations into the science of science can be subordinated: the investigator is interested in science either as a way to the knowledge of the world, or as a field of human culture. Let us call this first attitude the *epistemological point of view*, remembering, however, that we are concerned here with a wider range of problems than the one embraced by traditional epistemology. And let us call the second attitude the *anthropological point of view*, bearing in mind anthropology in that wide sense in which it is understood in the English-speaking countries (the science of man and his culture).

* First published as 'Nauka o nauce' in Polish journal *Nauka Polska,* Vol. XX (1935) No. 3.

The distinction of these two points of view is not a division of the problems: besides problems specially connected with the cognitive functions of science, and besides problems specially connected with that second attitude, there exist problems which may be approached from both sides. For example, the psychology of scientists may interest both those who study the connection of science with man's creativeness and the development of culture, as well as those who study the question of a scientific cognition of the world (what qualifications he who is to discover or prove new scientific truths has to possess?).

The interest taken in cognitive functions is very old. The history of investigations of this kind goes as far back as the times of Xenophanes. The interest taken in science as field of human culture is something new. It was partly derived from historical research, partly called forth by the development of modern sociology, and partly by practical needs (the question of the encouragement and organization of science). Research in this field is much younger than the science of religion, than the science of economic production, than the science of art.

Those interested in science from the epistemological point of view treat the existing scientific works and research methods applied by scientists only as examples and as material for conceptual analyses, for classifications, for considerations of the cognitive value of various possible activities and theses. The degree of importance, attributed from this point of view to the separate cognitive activities and fields of science, does not depend on its proper role in the development of human culture; the criteria of importance are established on the basis of the general postulates of the theory of knowledge and of methodology.

The theory of scientific cognition, trating its subject *sub specie aeterni*, may be in fact cultivated without a closer contact with the life of science.

In investigations about science, resulting from the interest taken in man and his culture, we have before us a concrete reality: the life of science with its triumphs and errors. And though we look forward to general formulations we must study this life of science in the fullness of historical circumstances.

Problems of the science of science

The problems concerning science in its manifold forms — both those already found in studies devoted to the science of science and those which have not yet been voiced — could be grouped according to various principles. They could, for example, be divided in three groups: the one taking into consideration all problems connected with the *personality* of the creative worker in science, the second pursuing all problems connected with the *activities* leading to the formation of science, the third concerned with the problems relating to science as a completed human *product*. This is one of the numerous possibilities. According to another, the same problems could be put under three other headings, analysing under the first all problems which answer the question what science is, under the second how science originates, and under the third — what effects it has.

While the principle of classyifying the problems quoted above only as examples — must lead to the separation of problems which hitherto were usually considered jointly, [2] the grouping given below takes into account the hitherto prevailing division of labour among the scientists. According to this proposition the problems concerning science would be grouped under five different headings, three of which would form the backbone of our new branch of science.

(a) Among the first of these three fundamental groups we would propose to count all those problems which hitherto have been usually treated by philosophers and which could be called the *philosophy of science.* Here would be placed such problems as that of the conception of science and the numerous controversies connected with this question, as to what is still, and what is no longer science. Here the classification of the different branches of science would find a place. Here would be included a large group of general methodological problems: the analysis of the way of proving various kinds of statements, the analysis of the role of 'as if' in scientific cognition, [3] analyses of such conceptions as that of scientific law, hypothesis and such like. In this group the epistemological point of view, spoken of at the beginning, would be dominant.

(b) The second division of research about the science of science could be called the *psychology of science*. Here will belong such problems as, for example, the problem concerning the psychic development of a scientific worker, the distinguishing of certain types of investigators, the psychological analysis of various types of research activities and of various stages of scientific activity, the question of capacities necessary for the cultivation of certain branches of learning etc.

(c) In the third group science does not appear as in the first one, detached from any historical background, but against the background of social life and of the whole of cultural life in general. This division, in which the point of view, called by us anthropological, predominates, could be given the name of the *sociology of science*.

Here will be included such problems as that of the relations between science and other products of culture, like art or religion. Here will be voiced the dependence of the development of science on economic conditions, on the structure of a given society, on the organization of education. Here not only the factors by which science is influenced, but also its effects on cultural life, its influence on the economic conditions, on law, morality etc. will be examined.

One group of these problems is at present being discussed by German sociologists in the so called '*Wissenssoziologie*'. [4] These sociologists are in vigorous opposition to the widely spread treatment of the evolution of science as if in this evolution only immanent factors played a decisive role. [5] Science like all other human products is in their opinion '*sozialbedingt*', and 'the problem of knowledge being socially conditioned in its various forms' is one of the central focuses of their interests. [6] In accordance with the meaning of the word '*Wissenssoziologie*' their investigations do not only embrace science. They are interested in knowledge in general. They are interested in the formation of public opinion, the arising of certain popular conceptions of life, characterizing a given epoch, the shaping of certain ideologies. Related to these studies was the article by F. Znaniecki, entitled 'The Subject Matter and Tasks of the Science of Knowledge' published in Vol. IV of *Nauka Polska*, 1923 (p. 1 of the present volume). Like the works mentioned before it had a rather programme-like character. As this is a new field of research it contains, of course, more plans than realizations.

(d) In close contact with the group of sociological problems, singled out by us a moment ago, remains a large group of problems of a *practical and organizing character*. These problems have been hitherto chiefly undertaken by institutions carrying on the promotion of science and using for certain practical ends the theoretical results of the former groups. Here belong such questions as the general problems of the organization of science, social and state policy in relation to science, the organization of higher institutions of learning, of research institutes and of scientific expeditions, protection of scientific workers etc. etc. These problems deserve to be singled out from the sociological group on account of their practical character and on account of the fact that in the hitherto prevailing division of labour they were allotted to somebody else.

(e) Finally the fifth and the last of all the groups of problems enumerated by us, consists of *historical problems*. The history of the conception of science, the history of the conception of the scientist, the history of the separate disciplines and of learning in general, these are materials which can be used by all the groups of problems, which have been mentioned above.

Undoubtedly these groups overlap in various ways, but perhaps there are in them fewer problems of a doubtful character than in other attempts at a classification of this diverse material. This attempts is of course only a provisional arrangement at which it would be difficult to stop. Looking at the question more closely, a further differentiation within the groups would prove indispensable, for the time being, however, we disregard this, sacrificing the accuracy of our presentation for its greater lucidity.

The set of problems proposed by us does not cover the same ground as those that have been previously classed by some under the name of 'science of science'. The history of German philosophy has for a long time made us familiar with the name '*Wissenschaftslehre*', but those who used it did so in a different sense from that in which it has been employed here.

Leaving aside Fichte who called his whole philosophical speculation by that name, this term was used in Germany chiefly to denote logic with general methodology, or logic with general methodology and the problems usually included in epistemology. In almost the same sense the term '*Wissenschaftslehre*' was used in the work of Bernard Bolzano, the subtitle of which explained '*Versuch einer ausführlichen und grössentheils neuen Darstellung der Logik*'. It was logic understood in a very wide sense, with which we were later made familiar by German textbooks of the end of the XIX century and the beginning of ours.

Limiting philosophy to problems of logic, methodology and epistemology, Prof. Kotarbiński in the quoted *Elementy...* [1] proposes to call philosophy the science of science. It is clear that the territory of the science of science, as spoken about in our article, does not correspond with the area marked out in these propositions. It is too wide an area in relation to ours, and at the same time too narrow. Only a part of the problems of logic and epistemology will go into our first division of the science of science, while all the remaining divisions singled out by us will be in relation to that 'science of science' something new. Closer to our problems is W. Schingnitz in his article entitled 'Scientiologie' (*Minerva Zeitschrift,* 1931, Hft. 5/6 and 7/8. Jahrg. 7), but his remarks, though frequently valuable, are fragmentary and do not compose a distinct whole. Problems related to those singled out by us in our psychological and sociological group, are also enumerated by E. Radl in an article entitled 'Zur Philosophie der Wissenschaftsgeschichte' (*Scientia,* 1933, Vol. LIV, N. CCLIX-11). They are alloted by him — with which it is hard to agree — to the historian of science. This research, according to him, should be conducted from the point of view, which we have called anthropological in the present article.

Schingnitz in the article just quoted did not only enumerate various possible types of research in the science of science, but also proposed a name for it: he wished to include them under the term of *scientiology*. Those who wish to replace the expression 'science of science' by a one-word term sounding international, in the belief that only after receiving such a name a given group of problems becomes officially dubbed an autonomous discipline, could be reminded of the name *mathesiology*, proposed long ago for similar purposes. [7] This term might satisfy those purists who in the term 'scientiology' may feel offended by its combined Latin and Greek origin.

Question of the autonomy of the science of science

The conception of a new branch of science usually becomes finally crystallized only when this discipline has gained admission to the university curricula. Before this takes place it has to overcome the doubts as to whether there are sufficient reasons to recognize in it a separate, autonomous branch. These doubts were raised also in relation to the 'science of science'.

From the theoretical point of view this is an insignificant matter. To a large extent it is a matter of convention whether we shall recognize a certain system of problems as a separate branch of learning, or whether we shall subordinate it to a more general science or assign it to several various branches. But as certain practical consequences are involved here, we shall for a moment consider whether in relation to the science of science there are more reasons for such doubts than in other fields.

Against the autonomy of the science of science the argument is raised that its problems already have their places in other formerly separated fields (psychology, sociology, logic with methodology, the theories of the separate sciences); the science of science would be therefore a discipline without any such problems which would be peculiar to itself. Let us, however, consider that the areas of the separate sciences are not in general marked out like political or administrative territories; within their confines condominia cannot be avoided. These condominia are usually widest in the recently separated branches of learning, because usually, before a new discipline can emerge, the problems which are to form its foundation already exist, but up till now have been assigned to other divisions. Therefore it seems quite natural, that in the first stages of the history of a new science the whole repertory of its problems can be taken from other fields. It is only in the course of time, when we have already become sufficiently familiar with the new frames, and when from the borrowed problems new ones arise, that we beigin to treat the problems of the new branch as peculiar to itself. What matters, is whether these various problems possess a sufficient internal unity.

This very internal unity of the science of science was also frequently questioned. The scope of research in science of science comprises investigations concerning widely different subjects; in some cases these will be into psychic phenomena, from which scientific works arise, or into the physiological conditions of scientific creativeness; then again certain social institutions are the subject of research, and still again the relations between scientific assertions, or between scientific assertions and reality etc. These various problems are solved by various methods. But also in this respect the science of science is in a similar position to many officially recognized disciplines. A complete uniformity of subject and a uniformity of methods can probably be found only in the aprioristic sciences. The link between the problems of one branch of science does not necessarily consist of the unity of methods and points of view. Such a link can be formed by the subject which is the centre of interest; and this certainly does not mean that this subject should be the only subject of research of the given branch; we are also right in speaking of the unity of a subject when heterogeneous objects are being investigated, so long as they are objects of research on account of their connections with that common centre of interest. Let us take general linguistics as an example. The student of general linguistics is concerned with psychic as well as with acoustic and physiological phenomena and also with the field of meanings. He is concerned both with the processes which take place in the mind and body of the speaker, hearer and reader, and with morphological and semasiological changes in the history of language, with the mutual interdependence of different languages and the peculiarities of various dialects of the same language. He applies extrospection as well as introspection, historical methods as well as those used by natural science. But the autonomy of general linguistics in never questioned, because all heterogeneous subjects which come into the scope of linguistic research remain in definite relations to *human speech* and therefore interest the student of linguistics.

The case is the same with the science of science. Here science is the centre of interest (it doesn't matter, whether we have in mind science as a product, or as certain activities). All other never entering the scope of research in science of science remain in this or that relation to science, and it is for this very reason that they are here the subject of research.

Another argument advanced against the autonomy of the science of science, is the lack, not so much of objective, as of personal unity. We have here in mind the doubt whether one man may successfully carry on research so manifold and requiring such a many-sided preparation; can there exist a specialist of the science of science, a specialist who would have to be experienced in general methodology, to be familiar with the achievements of psychology, of the theory and history of culture, etc.?

If we wish to decide whether these doubts hit at the autonomy of the science of science, we can again refer to the above-mentioned science of linguistics. The qualifications required to investigate thoroughly a given language are extremely manifold: phonetics, semantics or dialectology of a language are fields of research requiring a very diverse preparation. And what shall we say of one who wishes to become a specialist in general linguistics, where, besides the preparation in logic and psychology, besides historical and ethnological knowledge, it is indispensable to possess a knowledge of various languages which would make comparative studies possible.

The decision whether a certain group of problems can form a separate discipline, is not in fact dependent on whether one person is able to embrace all these problems. The scientific unities singled out in the traditional division of labour continue to live as unities, though, due to their growth, they have become impossible for a single individual to master. Such a principle of grouping the problems would take into account extremely casual factors: what one individual is able to master today, may be impossible for one to-morrow.

The contemporary life of science makes us more and more

familiar with group research, which, though it is done collectively, does not violate the unity of the discipline to which it is alloted. Such research is forced upon us especially where the same subject is to be investigated from various points of view, i.e., in situations in which our science of science is found. Its future will have to be entrusted to groups of people rather than to individuals.

Apart from this whole discussion we think that the legitimization of a new science should be brought about not by aprioristic considerations but by empirical ones. Only the pragmatic point of view should be decisive: is the given group of problems sufficiently vital for its importance to be emphasized by giving it a separate name? are there any concrete didactical considerations in its favour? have we any reason to expect fruitful consequences by grouping these problems in a common separate frame?

For if the question of the singling out of a certain group of problems into a separate discipline be insignificant from the theoretical point of view, it is not so from the practical one. A new grouping of problems lends additional importance to the original problems and gives rise to new ones and new ideas. The new grouping marks out the direction of new investigations; moreover, it may exercise an influence on university studies, the foundation of chairs, periodicals and societies.

Actuality of the science of science

Studies in culture, classed under the name of social or cultural anthropology are in general not concerned with science in the present sense of this term, because science in this sense is not a component of all the heterogeneous cultures with which the student of anthropology or ethnology is concerned. Almost every culture is characterized by some religion, some magic, art, technology, morals, but science in the perspective of time and space is a rather exceptional phenomenon. It plays, however, an extremely important role in one culture which is of special importance for us: in modern European culture.

The majority of contemporary sociologists think that it is owing to Science that modern culture is a completely new type in history, quite incomparable with all pre-scientific cultures. It has no more than three centuries behind it, and the gulf which divides it from all former cultures grows with every year as the influence of science on all spheres of life becomes more powerful.

The history of science goes back, of course, to earlier times. Certain characteristic features of our scientific culture had already had their precedents in history, be it in the Hellenistic or the Chinese or even the medieval cultures. There is, however, something entirely new in this uninterrupted development of science in the course of the last three centuries.

The realization of this fact will facilitate the proper appreciation of the science of science. Therefore without entering upon an analysis of the conception 'scientific culture' or 'culture based on science' it may be worth-while to glance at some of its features which are most significant and which have been much written about in publications of recent years.

(1) Pre-scientific cultures had to be stable in their foundations. Whatever had claims to complete stabilization met in them with highest approval. When, for example, religion was changed, it was either done in such a way that nobody distinctly realized it (slow transformation of religious tradition) or by means of a revolution when, instead of one 'unchangable' religion there was introduced another religion with claims to the same unchangeability. On the contrary, modern scientific culture is not only in a stage of constant changes, but this dynamic quality is accepted as a postulate by those who create it.

(2) An equally new feature of the scientific culture is its universality. There is only one scientific culture, absorbing all scientific achievements, wherever and by whomever they are attained. There are no competing scientific cultures, there are no competing sciences as there are competing religions or codes of law. All incongruity between various scientific theories is considered a provisional stage which has to be overcome in this or that direction. Our culture becomes also universal in the geographical sense: it has encircled the globe from pole to pole, on the waves of the wireless it penetrates the widest thickets of the jungle, it reaches the most isolated islands of the Pacific.

(3) Modern culture enables man in an ever growing degree to transform his environment according to his own aims instead of adjusting himself to it. In relation to former cultures this is a rather quantitative difference, but in such a powerful degree, that it is impossible not to see in it a qualitative difference as well. Not limiting itself to ruling over an extrahuman environment, science endeavours to assume control over human instincts, over social and economic forces. In the last few decades scientific plans have been forming for the organization of human life on a world scale. This was unknown to any pre-scientific culture.

(4) Changes in the conception of life, caused by modern science, are incomparably more far-reaching then the changes occurring at the transition from one pre-scientific culture into another. We have in mind here not only changes, directly called forth by certain new scientific discoveries, as, for example, the widening of the limits of time and space due to astronomical and geological discoveries, or the change of man's position in the world of living creatures, brought about by the discoveries of biology. Here are also involved certain general tendencies of thought. On the one hand the dynamic conception of human culture, which has been already mentioned, and on the other, rationalism. On the basis of scientific culture a conflict has arisen between science and the spheres of culture which have survived from the pre-scientific era. This conflict is presented in various ways: intellect and subconsciousness, rationalism and the traditional mental habits, science and religion. The future will show whether it will be a permanent characteristic of our new type of culture.

Scientific culture is only in a stage of formation. The speed of its development is growing, but its most characteristic features have not yet been wholly attained. At any rate the changes, brought about by science in social life and in human minds, suffice to lead us to the belief that the science of science will have to occupy a special place in the rank of the sciences of man and his products. It was brought into existence not only by new interests, but above all by the *new reality*. As the discovery of electricity brought with it the creation of new division of physics, so the development of science in modern society will bring about the rise of the science of science.

Practical application of the science of science

Besides theoretical interests in science, and besides its more and more powerful influence on the whole system of human life, *practical needs* also come into play. The growth of science requires an extremely wide and many-sided supplementary apparatus and the building of this apparatus requires theoretical studies. Practical needs, as we have mentioned, played an important role in the genesis of the science of science.

The practical applications of science of science do not require an explanation; the organization of scientific work, individual and collective, the organization of institutions, the protection of science by the state and by social organizations, the education of the scientist, all this — if it is to be fruitful — cannot to-day do without studies just as specialized and complicated as those which are required for the construction of large industrial establishments.

But the investigations of which we speak play yet another role, though not such a conspicuous and direct one: teaching what science is, contributing to form in the minds of scientific workers this or that conception of science, they at the same time influence their further creativeness; for science, like all other fields of culture, is a part of that particular sphere of reality whose history depends on what we think of it.

NOTES

[1] For this distinction see T. Kotarbiński: *Elementy teorii poznania, logiki formalnej i metodologii nauk*, Ist ed. Lwów 1927. Engl. ed. *Gnosiology. The Scientific Approach to the Theory of Knowledge*, Pergamon Press/Ossolineum, Oxford 1966, p. 311.

[2] So, for example, according to the first proposition the problems which we include in psychology, would be found both in the group of problems connected with the personality of the scientist and in that connected with the activities leading to the formation of science.

[3] We refer to the book of Hans Vaihinger, translated into English by C. K. Ogden under the title *The Philosophy of 'As if'*.

[4] See: K. Dunkman, 'Die soziologische Begründung der Wissenschaft', *Arch. für system. Philos. und Soziologie*, Bd. XXX, Hft. 1, 2; E. Grünwald, *Das Problem der Soziologie des Wissens*, Wien–Leipzig, Verl. von W. Braumüller, 1934; K. Mannheim, *Wissenssoziologie*, *Handwörterbuch der Soziologie*, A. Vierkandt, F. Enke, Stuttgart 1931; M. Scheler, *Versuche zu einer Soziologie des Wissens.* München–

Leipzig 1924; Al. v. Schelting, 'Zum Streit um die Wissensoziologie', *Archiv für Sozialwissenschaft und Sozialpolitik*, Bd. 62, Tübingen 1929; K. A. Wittfogel, 'Wissen und Gesellschaft', *Unter dem Bahnen des Marxismus*, Jahrg. V, Hft. 1.

[5] See the quoted article by Mannheim.

[6] See the quoted article by Wittfogel.

[7] This term was proposed by Ampère (see Eisler's Dictionary, article 'Wissenschaftslehre'.

Tadeusz Kotarbiński

A REVIEW OF QUESTIONS IN THE SCIENCE OF SCIENCE*

Translated by Christopher Kasparek

The chief task of epistemology is to acquire a knowledge of the world of the sciences that will facilitate the most effective contributions to the flowering of the sciences with the aid of social resources. It is useful for this purpose to distinguish questions concerning those philosophical aspects of the sciences having but remote connection with this task — e.g., questions of epistemological theory — from questions whose solution promises to be important to the task, e.g. questions involving the developmental dynamics of scientific research. It is especially important also from this standpoint to differentiate among the sciences by the degree to which they require materials and apparatus and by the degree of the anticipated utility of their achievements to technology and the economy, as well as by the degree to which their questions are connected with questions of an ideological nature.

The purpose of the present paper is to carry out a review of questions involving science that will make evident their relative importance to epistemology; while we see the essence of the latter in a study of the world of the sciences whose purpose is to prepare optimal assistance for scientific research by social institutions, particularly by institutions of public administration. [1] We will attempt, therefore to point out, on the one hand questions that are important to this end, and on the other questions that are,

* First published as 'Przegląd problemów nauk o nauce' in Polish journal *Zagadnienia Naukoznawstwa,* Vol. 1 (1965), Nos. 2–3.

to be sure, very interesting from the standpoint of a general view on reality but that are fairly unimportant to epistemology, questions which epistemology may with no loss omit from its repertoire. [2] The exposition and elimination of these questions will predominate in the first part of the discussion.

The better to approach the task at hand, it will be well to come to an agreement concerning the meaning of terms that we shall use in our arguments. The chief term, 'science', is fraught with a fourfold ambiguity. For it [i.e., the Polish *nauka*—Translator] means research of a stated kind, or the process of teaching, or that of learning, or finally very often designates systems of duly justified propositions. [3] Therefore we stipulate that in the present discussion we will be interested directly in only two of the above meanings, the first and the last: in 'science' understood as 'research', and in 'science' understood as the actual fruit of that research, that is, as a certain 'system of propositions'. We shall concern ourselves with matters of teaching, as well as with questions pertaining to learning, only indirectly, insofar as they prove useful in elucidating the latter questions.

There arises a dilemma of what to begin with: with science as research activity, or science as system of theses expressed in spoken or written words. To be sure, the authorities lend their assistance to the activity rather than to systems of theorems, and consequently science as activity would deserve, as it were, pride of place before science understood as the contents of learned books; nevertheless, that primacy pertains only to the dominant direction of our interests in the present deliberations, not to the rational order of the undertaking. And from the latter point of view the doctrines, theories, hypotheses and their interrelations come to the fore, since the variety of forms of research activity, and especially the variety of their needs, beome fully comprehensible only by reference to these intended or actual achievements of the research.

And so, at the very threshold of inquiry, an inescapable problem demands a moment's thought: what actually is science as system of propositions? And that problem divides into two sub-problems: what is a system, and what are propositions, the elements of the scientific system? Let us begin with the elements.

Some will say, "What else could, say, the propositions in a geometry textbook be, if not sufficiently justified statements?" Propositions, according to those who think this way, are statements — series of printers' symbols or of human speech sounds corresponding to these. Others reject this view. "Science", they cry, "does not consist of bits of printers' ink nor of their acoustic correlates: propositions are only symbols of thoughts, and it is precisely of thoughts that, for instance, Euclid's system of geometry consists. Thus it is thoughts — thoughts worthy of distinction — and nothing else, that constitute the axioms, lemmas and theses of his science".

"How's that?" we hear from yet another quarter, "geometry is supposed to consist of someone's thoughts, in other words, of someone's psychic acts? Any psychic act is someone's personal experience, and when a teacher has uttered a proposition and a pupil has understood it, then certainly two different events have occurred: the teacher's thought and the pupil's thought, but it would be absurd to say that two propositions have come into being. We are dealing here with a single proposition to which both have made reference, and of course with something that does not occur in time, since it would be patently absurd to say, for instance, that the proposition concerning the equality of the angles of an equilateral triangle — that this geometrical truth occurs as a process in time, as a process in someone's experience". Here the proponent of a fourth tendency comes forward with an explanation: "We distinguish", he says, "thoughts (or perhaps better, ideas) in the psychological understanding, and ideas in the logical understanding. The former are psychical processes in the egos of particular persons. It is not of such elements that scientific systems consist. On the contrary, they consist of the latter, of ideas in the logical sense. And what is an idea in the logical sense? It is nothing other than the sense of a declarative sentence — the meaning of such a sentence. And when the teacher utters a proposition, and the pupil understands it, then both are aware, each by his own act of thinking, of the same sense of the given sentence, and the proposition that both are concerned with is neither the sentence itself (whether graphic or acoustic) nor either thought of these persons, but that which both have in mind, that is, the meaning of the sentence". Thus, according to proponents of the latter view, science as a system of propositions consists of the meanings of sentences. And here normally a controversy bursts forth, often conducted not without recourse to 'name-calling'. Supporters of this view are called idealists, if only because — to the question of where and when those meanings of sentences exist — they reply that this question is no more sensible than, e.g., the question of whether integral numbers persist in a sleeping or a waking state. Consequently persons who think this way are accused of assuming the existence of objects that are only thought of — of objects that are 'ideal', not real, since they are located neither in time nor in space, do not act physically upon other objects, and are not subject to the physical action of other objects. The incriminated kind of idealism is also, with the perversity proper to traditional philosophical terminology, called Platonic 'realism', inasmuch as Plato is supposed to have assumed the existence of so-called ideas of a supreme kind, understood at any rate as extra-temporal and extra-spacial entities existing neither where nor when tangible bodies exist. But there also exists, besides the Platonic 'realism', a realism called Aristotelian, and this characterizes the fifth view in our register concerning the nature of propositions as components of scientific systems. Philosophers so oriented have been saying for centuries that science is interested in the general, but that the general dwells within things and does not somehow exist apart from them. In the controversy the historically given object of interest was the meanings, the senses of general terms. An analogous attitude toward propositions understood as the meanings of sentences leads — we feel — to their identification with certain relations obtaining in the world of temporal and spatial things. What, for instance, from this standpoint, is the meaning of the proposition in physics concerning the greater speed of light than of sound? It is a relation — the relation between the speed of propagation of light waves and the speed of propagation of sound waves. What is the meaning of the historical thesis that the monarchy of Alexander of Macedon distintegrated following his death into several independent kingdoms in which commanders of military units, who had been subordinate to him during his lifetime, became the kings? Clearly, the meaning of this proposition is a fabric of relations obtaining among contemporary and local people and things — of relations contemporary and local just as were the people and the things. Having heard this argument, one of the participants in the discussion is ready to ask, "If a scientific system consists in the meanings of propositions, and the meanings of propositions consist in relations among things, then does it not follow from these premises that a science concerning some field of reality consists in something that forms part of that reality, provided only that the relations form part of the things among which they obtain?" Finally, dissatisfied with all the foregoing solutions, a sympathizer with so-called nominalism — whose exponents for several hundred years have not ceased calling for liberation from the sway of nouns — will speak up. People generally feel that since a certain noun prowls the language, there must certainly exist an object designated by that noun. Consequently they attempt to describe this putative object and encounter difficulties, understandable to a critic, whenever they hit upon a vacuum. It is the same with the problem under consideration. We have the nouns 'science', 'system', 'proposition', 'idea', 'thesis', etc., and so we at once posit the existence of objects ostensibly designated by these — the existence of their supposed designates. And that leads to a fruitless search for a satisfactory answer to questions of the kind considered above: e.g., whether propositions are series of letters printed on paper or something altogether different, and in that case — what, etc. The adherent of nominalism, however, looks at all this suspiciously and proposes the following way out of the verbal tangle. To be sure, one says that there exists a Science with a capital S as a collection of sciences, that there exist individual sciences as systems of propositions, and the propositions themselves, the elements of the systems. But such statements are only shorthand for statements that are free of nouns used in the shorthand. What is really meant is that different people study different things, try to and do find out what those things are like, and tell other people that they are this way and not the other. To that end they utter or write down propositions characterizing them as knowing that such and such things are thus and so. Some study certain stated things, others other things, some ask how a given thing is built, others ask from what other things this thing arose, etc. And that is all; there are no objects such as 'sciences', 'propositions', 'systems', there are only things — things studied, e.g., inanimate physical bodies, plants, sentient organisms, people, human collectives and so on, and among other things — things that investigate, that is, once again people — and, beyond that, things that are products of the things that investigate: namely, on one hand informed people, and on the other, manuscripts, books, and in these, components of them that are also things, the strips of little nubs of ink called propositions. But these strips of nubbles are not only coloured thus-and-so, shaped thus-and-so, deployed thus-and-so: they also mean something, they express people as thinking that things are thus and not otherwise. What we are dealing with here is something similar to coins or banknotes. A hundred-złoty note is a thing, since it is a physical body, but it is not only a flat, rectangular, easily combustible thing of paper, but — what is usually most important — it is worth so much, because for instance it is enough to give it to the right persons seated at the right counter in order to be insured for a month against loss of health, i.e., in order to become justifiably certain that if one falls ill, physicians will attempt to help. And the comparison of propositions with banknotes will be very useful in the present discussion. In what way, must be explained.

The author, though inclined to sympathize with the nominalists, does not intend to tilt here for the victory of their way of viewing the world. The task at hand is only a review of questions in the sciences of science, and — difficult as it is to characterize questions without also showing certain possible ways to their solution — offering answers to the questions plays the role of an ancillary operation and would be crucial only if without adopting a stated answer it were not possible to advance the review of questions in the sciences of science. Subsequent arguments ought to show that in our case this is not so. Moreover, and more importantly, our intent is to prepare guidelines for instrumentalities that organize and support scientific investigations. And these organizing and sponsoring tasks do not depend on previous resolution of the question of who is right in the dispute over the nature of propositions — nominalism or one of the other philosophical schools; much as questions of financial policy, e.g., the struggle against inflation, do not require for their solution the prior assumption of a position in the field, if one may say so, of hundred-zloty philosophy. Thus if we have taken the occasion to expatiate on the difficulties touched on above, it is because the questions in the sciences of science include problems, a familiarity with which seems to us important for purposes other than those of organizing and conducting research, e.g., for purposes of a general education. For the latter includes, among other things, an awareness of the role of language in constructing a picture of reality — a picture that is at once a likeness and a deformation of that reality. Nevertheless, we stipulate that the epistemologist (if his task be understood as decided on at the beginning) has a right to operate with the terms 'proposition', 'science', etc. as blithely in this respect as do practical financiers in operating with the terms 'legal tender', 'hard cash', 'bearer check', and so on, without wondering whether money as such is or is not tangible, etc.

This does not mean, however, that in our review of questions we have freed ourselves of purely contemplative problems, for next to the question of the nature of propositions a star of the first magnitude shines in the firmament of the science of science: the problem of the nature of truth, more precisely, the question of what it means that a given proposition is true. And about this centre of intellectual gravity there circle planets in the form of derivative questions, among which — distinguished by the importance of their content — are questions clustered about the concept of the relativity of true propositions. In the realm of that prime question a continually renewed vitality is shown by invectives against the classical interpretation of truth. Nevertheless, the rival approach of finding the essence of truth in utility of one kind or another is somehow ceasing to be persuasive to penetrating minds. They continue to accept utility, interpreted in one way or another, only in the role of a certain — whether it be chief or ancilllary — criterion of truth. Truth understood classically is king of the castle, and that means truth understood as agreement with reality. Some prefer to say that a true thesis is a reconstruction of reality, others — and their voice for many today rings authoritatively — have selected as their battle cry the term 'reflection'. Behind these verbal differences lie different varieties of a common view: an object about which we state something is in every case a certain way or bears a certain relation to other objects; and we state truly that which we state, if it is just the way, or bears just the relation to other objects, that we say it is or does. The above-mentioned varieties of the classical concept are inconsequential, secondary shades of the same colour. Great differences arise, however, whenever philosophers have occasion to analyze the concept of reality, whose true thesis is supposed to be correlate, reflection, reconstruction, etc., depending on the predilection of particular varieties of the classical attitude. To materialists, who on this point are in accord with the views of the generality of sensible practical people, the objects experienced are the way they are and not otherwise prior to and independently of what anyone thinks of them (except, of course, for products of human activity that are the fruit of concepts earlier entertained, and so on). This is denied, however, by idealists of several different denominations who either appeal (erroneously, it would seem) to the view that all objects given to us by experience are complexes of qualities (colours, shapes, masses, tastes, etc.) that patently do not exist prior to and independently of the experiencing, or who draw (illusory, it seems) consequences from the fact that it is impossible, on pain of contradiction, to experience an object that has not been experienced by someone, and that every time someone experiences something, the object experienced has some form determined precisely by the fact and manner of the experiencing.

And now another question: is this entire matter of the nature of the truth of propositions of sufficient consequence to the proper organization and sponsorship of scientific research? We entertain serious doubts as to that. In order to justify them, perhaps we will be permitted to make resort to the teachings of the fabulists. They say that a certain philosopher fell into a deep crevice that he could not get himself out of; but when a rope was dropped down to him, instead of seizing the instrument of deliverance, he proceeded to deliberate over the nature of a rope which, in his situation was the height of irrationality. In the social sponsorship of the sciences let us operate with the colloquial concept of truth whenever that appears necessary. That will suffice.

We may well, however, have to judge differently the utility, to the stewards of the sciences, of the controversy over the relative or absolute nature of truths, and hence also of true propositions in science. Wishing, however, to present this controversy, or rather this tangle of controversies, in outline, it is difficult not to emphasize at the outset that this very matter makes particularly evident the necessity of preceding analysis by a recognition of the ambiguity of the expressions involved. When someone states that a given truth or any truth, truth in general or truth as such, is 'relative' and not 'absolute', he may mean, for instance, either that what is true is so depending on who utters the given proposition or depending on the circumstances in which he utters it; or that the truth of a thesis changes depending on time, so that something that was the truth yesterday may cease to be the truth tomorrow or the other way around; or that a given proposition contains a partial, incomplete knowledge that does not exhaust that which might be known about its object upon further study, or a knowledge only approximate and imprecisely depicting reality, or finally an uncertain knowledge requiring a better grounding in order that we might depend upon it without risk of committing a mistake. Discussions concerning the relativity of truths have elucidated the question to such a degree that, for persons professionally acquainted with it, little remains to discuss, if one knows well in each case of a reborn controversy what sense of the term 'relative truth' designates the terrain of the controversy. Agreement prevails among the experts that it cannot be the case that a thesis with an unambiguously defined meaning can be true or untrue depending on the persons voicing it or on the circumstances of its voicing; the only thing that remains debatable is whether certain predictive propostions, that is, propositions expressing conjectures concerning future facts, are distinguished by the circumstance that they will become true or false only when a particular agent acts in a particular way, at his own personal discretion. There are, however, no serious differences of opinion as to the progress of knowledge, which by the nature of things leads from propositions more fragmentary, less precise and less well founded, to propositions that consider more sides and aspects of the object, represent it more precisely and are more solidly founded. And this fact confutes the tendency to turn away in contempt or indifference from scientific workshops that have so far given forth only very fragmentary results, far from precision and not meriting full confidence. For it is from such inescapably provisional attainments that the path of rational results leads to achievements of a vast scale, striking precision and remarkable — at times even an unshakable — durability.

With that we conclude our review of questions involving propositions, the components of the sciences understood as systems of propositions. That removes us from the domain of the theory of knowledge, in which we have tarried as we have pondered the nature of that to which truth is properly ascribed and the nature of truth and its relativity understood in one way or another. But we are not yet abandoning the domain of investigation concerning the sciences as contrivances — as systems of propositions. And since a scientific system — at a given stage of evolution of a given science — is in every case a collection of propositions that are at that stage adequately justified, consequently there now come to the forefront questions from the domain of the justification of propositions, that is, from the domain of logic. Do we thereby enter the domain of the psychology of thought? Hardly. Several decades ago, it must be admitted, the psychologists felt very confident in reasoning as follows: logic is a science of thinking, and thinking is a psychic phenomenon; and since all psychic phenomena are objects of study for psychology, hence logic must be part of psychology. In their interpretation, for example, the law of non-contradiction, in logic, would be a characteristic of attentive thinking and would state that no one, in thinking attentively, experiences simultaneously two propositions, one of which would constitute a contradiction of the other. However, in the heyday of psychologism there quietly matured and finally burst out a loud protest against so evident a confusion of concepts. "How's that?" exclaimed the antipsychologists, "does, for example, the thesis of the categorical syllogism describe some typically occurring thought process, does it state some natural regularity in the temporal succession of its phases, is it subject to justifying argumentation in the form of an inductive generalization on the basis of compiled particular observations — indeed, is anything at all said in it about mental acts?" We find nothing of the sort in the wording of this thesis, which reads: if every *S* is *M*, and every *M* is *P*, then every *S* is *P*. Today the battle between psychologism and antipsychologism, in the eyes of virtually all the experts on the subject, is properly regarded as being concluded in favour of the latter camp, and in the course and as a result of the controversy it has become apparent that the nature of propositions in logic is not only alien to the nature of psychological knowledge, but shows striking similarity to the truths of mathematics, especially to elementary truths in set theory and in certain kinds of algebra. True, there is a connection between logic's theses and thinking, but not such that they might describe the actual course of certain thought processes. The logician does not ask how one thinks but how one ought to think if one wants to draw valid conclusions. And in order to answer that, he considers how the constructed propositions are deducible from propositions of a stated structure. That relation in turn is designated, for example, by what relations among sets discussed in these propositions condition in suffcient fashion other relations among the sets. And the mentioned law of the syllogism turns out to be marked by the fact that the relation of a set being contained within a set is a transitive relation. In other words, by the fact that, if the first set is contained in the second, and the second in the third, then the first is contained in the third... Nothing remains here of the semblance of the psychologicity of logical propositions.

However, before having cut ourselves free of psychologism in logic, we pull further on the thread of questions of a logical nature, we ought to devote a word to questions, prospering fairly well entirely outside the realm of logic, of the psychology of science. Involved here are matters such as the development, during the individual's maturation process, of the concepts with which the so-called 'exact sciences' operate, such as forms of intuitive determination of causality, such as the importance of so-called 'elemental cerebration' (that is, the brain's spontaneous activity during sleep) in the shaping of concepts, etc. [4] And next to the psychology of science, meshed inextricably with the general psychology of thinking, we may set the sociology of science, practiced systematically since recent times, whose questions tie in inescapably with questions of the theory of organization of scientific research — about which below, under the rubric of methodology. And the results of investigations both in the psychology of science and in the sociology of science cannot but interest the epistemologist as such. A psychological awareness is important, if only in connection with efforts in the proper basic education and continuing education of investigators, and sociological generalizations will be helpful in the rationalization of team research. [5]

The author would be gratified if he were to obtain readers' consent to the following statements and directives. The practice of logic, the science of justifying theses, is a separate specialty. Its particular questions do not enter the purview of the specialty of the epistemologist as such. But the epistemologist, like any other man of science, must in his calling operate with the logical bond, must reason efficiently. Some are capable of doing this without special study of logic, but such study may be of use to the epistemologist, inasmuch as it may be of use to any investigator or thinker. That is not the end of it, though. In order to be able to deal rationally with the sciences as socio-historical creations, one must understand the internal structure of the sciences, the dynamic of its development as the sciences progress. And since the logical bond is a significant element in the structure of the sciences, it is difficult to exaggerate the utility of a logical training for all those who desire to work successfully in the field of epistemology.

Let us repeat: an awareness of the actual nature of the logical bond forms an indispensable introduction for grasping the structure of scientific systems. These scientific systems are wholes whose component propositions are bound together by relations of a logical nature, and dominant among these relations is that of deducibility. This is not the place to list all the types of scientific systems known to the history of the sciences. A different structure is to be found, e.g., in Euclid's system of geometry, Newton's mechanics of bodies, Mendeleyev's periodic system of chemical elements, Linnaeus' classification of plant species, Mommsen's genealogy of the evolution of the Roman state, etc. What is certain, however, is that modern mathematics has worked out the stereotype of a scientific system with the logically most perfect structure, and that system has consequently become one of the most clearly distinguished objects of interest for the science of science. What I have in mind is the concept of a formal deductive system. In such a system the proof of a given proposition consists in demonstrating its obtainability by way of reshaping the external structure of specified axioms or of previously proven theses. It is assumed that the logical bond is sufficiently determined by specified transpositions of propositions and of certain components of propositions, given established meanings of copulas linking these parts and of other stated linking elements. A deductive system may be used once as a statement by a given science about certain things, and another time as a statement by a different science about other things, provided only that in both fields there prevails a sufficient analogy in systems of relations, that both fields are, as it were, models of the same formal scheme. A good example here will be the formal system known as Boolean algebra, which can be interpreted either as a theory of relations among the ranges of terms, or as a theory of logical relations among complex propositions binding them independently of the internal structure of their component propositions, or as a theory of the structure of particular kinds of electrical circuits. In recent decades, in-depth studies have been conducted into the conditions that a mature formal deductive system should satisfy. Involved here, among other things, are non-contradiction in the system of axioms, the mutual independence of its components, and its completeness, that is, the possibility of proving on its basis a positive or negative answer to any properly formulated question in the field covered by its terms. These investigations constitute at present the acme of studies in the sciences of science, understood as product, as collection of propositions bound together by relations of a logical nature. We believe that the epistemologist, as such, ought to be aware of the vitality of the tendency toward formalization of the sciences, of the effort, manifest in the various sciences, at modelling themselves on the schema of a perfect formal system, and for that reason the cardinal questions of the logical bond and of the structure of a deductive system based on it must be obligatory for him. Formalization is one of the forms of the mathematical tendency in the development of scientific systems. Other forms of it are, on one hand, systems — bound up with the concept of probability — of propositions concerning statistically palpable relations, and on the other, simply the passage of various sciences from the phase of ascertaining qualitative relations to that of studying quantitative relations, which is connected with use of the concept of measurement in its sometimes exceedingly sophisticated and abstruse forms. Measurement in turn supplies quantitatively expressed experimental data, and the concept of final experimental data expressible in so-called protocol propositions has become one of the chief objects of contemporary study in scientific systems, along with the concept of the analytical proposition, that is, one whose truth is guaranteed by its very structure and by the very meanings of its component parts. Analytical propositions dominate in mathematics, protocol propositions lie at the foundations of the natural and social sciences. The epistemologist, as such, must be oriented in the evolutionary traditions of science, and these evolutionary traditions cannot be understood by anyone who is totally unfamiliar with the questions, just touched upon, of the diversity of scientific systems.

In accordance with the distinction formulated at the beginning, however, the term 'science' may be understood not necessarily statically — as system of propositions — but also dynamically, as complex of operations comprising research activity with respect to one object of research or another. Science so understood involves many questions of which we shall speak now.

The vast body of questions in the history of the sciences contains a truly inexhaustible variety. When we utter these words, however, doubt at once springs up as to whether we are addressing questions pertaining to an activity. Ready to protest against such an interpretation are those historians of the sciences who take a similar view of the object of their investigations as do, for instance, historians of literature or historians of art. From those camps one hears a warning not to confuse the objects of their investigations with the activities that were the cause of those objects coming into being. The historian of novels is not concerned with the biographical relations comprising a description of the vicissitudes of the author of this or that novel, and likewise the historiographer of painting, if he keeps to his proper task, is far from deliberations over the behaviour of the creators of forms of one sort or another characteristic of a given type of work in the visual arts. The former is concerned to discern the content and form of works that are novels, to characterize individual novels, to trace the direction of their evolution over time; the latter is interested in the variety of the pictures themselves and in the trend of changes in their pictorial content with the passage of the ages. Similarly, the history of science would have for its task reconstructing the history of the sciences as products, showing their content and structure as well as the changes in their content and form. Ascertaining the relation of a later phase of a given branch of knowledge to subsequent phases, the evolutionary trends that manifest themselves in the history of the sciences — these are among the tasks of the historian of science according to those who think this way. But neither the motives that encouraged individual investigators to their investigations nor the way in which they developed their inventiveness, nor anything at all of what belongs, so to speak, to the 'kitchen' from which the theories emerged, would belong to the tasks of the historian of the sciences in such a construction of his calling. But in that case, in turning our thoughts to problems of the history of the sciences, we would not be entering the terrain of research pertaining to science understood dynamically; on the contrary, we would be continuing on in the sphere of deliberations upon the sciences understood statically, that is, understood as systems of theses, only in the aspect of the temporal succession and maturing of doctrines over time.

Nothing, however, constrains historiographers of science to such a limitation in their objectives. On the contrary, the investigators' workshop itself, the active struggle of the investigators against adversities and resistance, their attempts, their experimenting, their spinning of conjectures, their entanglements in nets of errors, the drawing of ideas out of obscurity into the light, the whole promethean drama that has been playing for centuries in the great forum of history likewise has a right to be an object of interest to historians of science and is in fact an object of inquiry for many of them. [6] Furthermore, the collaborative linkage of research results with creative efforts of other intent, e.g., with technological, economic and other efforts, and the constant pressures of the economic base on the intellectual superstructure and, vice versa, those of the latter on the former — all this is part of the program of the history of the sciences thus understood. [7] And that is why the questions pursued by the history of the sciences clearly have importance in preparing a rational distribution of resources to the workshops where theories — complexes of propositions — are, so to speak, fabricated. For the history of the sciences teaches us about recurring relations, for example about how results of utilitarian importance have sprung from purely cognitive interests [8] and how often research in a given field has yielded results unexpectedly useful to research in fields seemingly vastly remote from it. What is more, progress in a given activity can be assisted consciously and effectively only on condition that one first acquires an orientation in the evolutionary tendency of its spontaneous momentum. And about that only a deeper retrospection (especially of the most recent, as it were current, history of science that is *in statu nascendi*) can supply the indispensable information. [9] In fine, it is also true that the contemporary investigator can learn not a little from the old masters of method. And in the social-practical sciences such as ethics, pedagogy, jurisprudence and others, the works of the old masters remain a continuing great source of valuable ideas.

Scientific-investigative activity is growing and with all certainty will continue to grow. Hence the interest in the questions, what are the characteristics of this growth, what are its prospects, and what sort of interventions are required of the organs that watch over its advance? The following generalizations have become, it seems, indisputable assertions. In every field of research the wealth of achievements is growing at such a pace that ever more effort is required to adequately master the knowledge already established, sufficiently to be able to push it further.[10] Making itself felt is the necessity for an ever more rigid specialization and the accompanying need for remedies against an ever more threatening restriction of the specialist's mental horizons. And ever broader spheres of research tasks are being embraced by the necessity of organic cooperation among specialists;[11] the postulate of synthesis, of integration equally in the work of the individual scientist and in the activity of teams, is becoming imperative. The accumulation of an inexhaustible mass of publications of scientific content renders indispensable an information service of ever better quality, cut to ever greater measure. The vast and ever growing dependence of technical and economic progress on the achievements of sciences and the analogous dependence of the advances in science on technical apparatus, give rise to numerous problems. Ever more apparent is the need for planning scientific research — among other things in connection with the need for long-range budgeting, and at the same time there is no shortage of symptoms of the need for reasonable moderation in details of planning research questions and in adjusting the style of planning to the peculiar properties of scientific research activity as such. Theoreticians devoted to their investigations are finding, to their dismay, that administrative functions — multiplying and becoming complicated together with the growth of research establishments—are absorbing more and more of their time and energy. Those are a handful of examples of evolutionary tendencies in science connected with its growth and permitting anticipation of the directions of indispensable efforts that will have to be undertaken on at least a national scale in order to satisfy the needs of science. More than one such objective will require international agreement since (and this is part of the established historical synthesis) science is becoming in ever greater degree the common concern of humanity. Thus it is little wonder that an international congress of historians of science was planned for 1965 on the theme of what the science historian can say about the future of science on the basis of scientific knowledge about, the previously traveled routes of its history.

This is the place to raise a dramatic problem that will not bear neglect. Do the advances of science bring more benefit than harm, and if they bring rather more harm (the atom bomb, etc.), then would it not be sensible to slow the growing tempo of its poisonous blossoming? This is a serious question — no invention of embittered visionaries. Nevertheless, we continue without hesitation the present discussion, which aspires to be useful to the positive appraisal of the need, and possible forms for support of science by the authorities. For we believe that only the blossoming of science can radically curb the dangerous side effects of its progress. It is probably also worth recalling that there is as yet — regrettably — no sovereign organization of humanity; and in an age of — at times, sharply antagonistic — competition, a society with incompatible aims, a society that neglected the development of science, would have to lapse into a humiliating dependence on other societies.

The history of science is surely, above all, the story of how science was created. And since the investigation of how something was done is in great part the study of method, consequently in pursuing the history of the sciences one of necessity enters the domain of methodology.

Here surely, if anywhere, the principle terms require definition. We have in mind the very term 'methodology' and the term 'method', from which the latter derives. [12] By method is generally understood a way of acting, consciously adopted and regarded as being repeatable. Corresponding to this interpretation is the construction of methodology as a science about just such ways of acting. Hence we sometimes notice that an author will term 'general methodology' the field of research, by others called 'praxiology', that studies ways of acting with the objective of devising the most effective ways. The adjective 'general' is added to the noun 'methodology' in order to distinguish the study of those ways that are useful in every field of human endeavour (i.e., husbandry, healing, military defense, administration, jurisprudence, education, etc.) from those that are peculiar to the work of the scientist as such. In order, however, to distinguish the latter, more special domain of study, we often resort not to the device of dropping the adjective 'general', but rather to limiting its scope by creating the term 'general methodology of the sciences'. This discipline is concerned especially, among other things, with ways of reasoning, such as deduction, reduction and their variants — inference, verification, induction, etc. — as well as with operations of conceptual clarification, classification, and so on. But why does the bipartite term 'scientific methodology' not suffice to denote it? Answer: because scientific methodology is in turn itself either general or particular; general methodology studies modes of scientific procedure that are valid over the entire domain of Science with a capital S, whereas particular methodology is the set of particular methodologies, each of which is interested in modes of research peculiar to a given kind of science, e.g., in experimental procedure in the physical and biological sciences, in introspection or testing techniques in psychology, in so-called textual criticism in the philological disciplines, etc.

Neither the term 'method', however, nor the term 'methodology' are always used in conformity with the conventions summarized above. By method is often understood, for example, the set of general assumptions adopted as a framework or as guidelines for research; for instance, whenever the advantages of Marxist methodology are discussed, what is — at least usually — meant is that research conducted in Marxist fashion is unburdened either by fictitious supernatural entities or by the illusions of Platonic realism in conceptual questions; that it is based on the assumption that 'existence determines consciousness', that truth is a 'reflection' of a reality that exist independently of perceiving subjects, etc. We believe that the differences in interpreting the terms discussed, as touched on here, must constantly be kept in mind in the frequent discussions revolving around the term 'methodology'. They are of importance in resolving practical questions and find expression in support or restraint of certain research undertakings. We must zealously strive to prevent decisions in such matters being based on misunderstandings.

Since a moment's attention has been devoted to the term 'scientific methodology', it will not be amiss to concern ourselves for a moment, once more, with the term 'science' itself, in order to elucidate the question of its scope, with science being understood as research. When do we 'do' science, as opposed to other activities? It is worth asking that: for there are differences of opinion in the matter. For instance, one hears the view that any scientific work boils down to discovering the truths that objectively govern reality and to justifying the theses that state them, and that the pursuit of history, for example, is scientific work to the extent that it seeks to discover regularities not discovered heretofore or to construct proof of their existence. Given such an assumption, any historical research limited to characterizing the peculiar features of individual past events or to connecting them in evolutionary series, or limited to genetic or causal interpretation of the past, would have no right to be called scientific work, just as the investigations of the police, undertaken in order to bring to light the culprits and the motives behind a crime, do not. Professional historians do not, however, consent to such a restriction of the concept of science. It was from their quarter that the awareness emerged that not only those research efforts are science, which set for themselves nomological tasks (determination of the laws of existence) but also those whose aims have an idiographic character (the distinguishing, arraying and interpretation of individual facts), provided only that the facts studied merit attention due to their importance in piecing together history.

But the supporters of the fairly narrow interpretation of science mentioned above are fencing themselves off not only from idiographs; they are attempting to build a wall as well on the other side of their own stamping grounds, by denying the name of science to the designing of constructions that are needed for something, be it machinery or plans for collecive action or legal systems, etc. For science, in seeking the laws of existence, is held to study always that which is, never that which might be done or which ought to be. But here again numerous representatives of modern technology refuse to agree with this. "How's that", they cry out, "designing a mathematical machine is not scientific work? Concepts for setting up a nuclear power generator are not scientific ideas?" If the fact that designers of technical apparatus do not study that which is — do not study reality — is an argument in favour of the opposition view, then this argument does not suffice to defend their position, since it is simply wrong. Physics too — alongside of mathematics, the most representative discipline in respect of its scientific properties — does not content itself with describing the persistent consequences of stated events but asks, for instance, how the motion of a body would occur in a friction-free environment or in a space free of any action from without. Thus it inquires into the consequences of a fictional set of circumstances, without thereby ceasing to study that which is, namely, what events constitute indispensable conditions for a stated fictional set of circumstances — it studies that objective relation among them. Similarly, a designer of machinery inquires into the conditions that would have to be satisfied by apparatus that fulfilled a stated set of assumptions and conditions, and in this sense he remains a student of reality, without ceasing to be in his further objectives the creator of new instruments. And our engineer-scientist's retort to his adversaries, in defending the scientific character of technical thought, may serve as model for a possible, equally just rebuttal by social engineers, directed against methodologists who deny the scientific nature of social engineering: with this difference, that in the justification of social programs, emotional value judgments are also voiced, especially positive judgments of that which ought to be and negative judgments of that which ought not to be. But these are not regarded as truths. "Why not?" ask sociotechnical thinkers. "If it is true that snow is white and sugar is sweet, then why should it not be true, for example, that fraud is disgusting and public confidence is a value worthy of pursuit?"

The French and the English like to use the term "sciences" to denote the complex of the natural sciences and mathematics, thereby narrowing the interpretation of science to a relatively limited field, albeit one not identical with that to which the opponents of the scientific nature of idiographic history and technical design would like to restrict it. Among the Germans, 'science' is '*Wissenschaft*', and *Wissenschaft* embraces both jointly. And we must concede that a term thus broadly construed is needed, in order to verbalize their activities, by the institutions that supply scientific facilities with all that they need. Therefore we propose on our part to understand by science simply any investigative discipline that is sufficiently developed to merit cultivation as an object of instruction at institutions of higher learning.

Among problems of a methodological nature pertaining to science as research, we should mention above all the problems of general methodology, that is, praxiology, which have application as well to the world of learning, inasmuch as their applicability encompasses the entire realm of human activities. It is, however, difficult not to observe that we have sought to touch upon virtually all that is fairly important here in summarizing the evolutionary tendencies in scientific advance that are manifest historically in connection with the growth of the sciences. Some examples: how are we to conduct ourselves in face of the growing collectivization of research, of its growing specialization, the growing intermediacy in the relation between investigating subject and the object of research (access to the object via a mounting accumulation of achievements and through the intermediary of ever more sophisticated and ever more difficult to master apparatus), etc. There was mention further of the tasks of a developed reporting system and of documentation generally as well as of planning (which also require supervision, as well as the organization of a perfected communication among fellow investigators of objects)... These reminders will surely suffice to show the vitality of the general-methodological way of looking at things also in the fostering of the sciences. But we shall add that full applicability is to be found here for the categories of preparation, simplification and concentration of operations and all the other praxiological categories, and that each of these contains within it a wealth of important questions.

No less rich is the repertoire of problems that characterize the general methodology of the sciences. Examples of matters that preocccupy contemporary methodologists include: Liberation from badly posed questions and the attainment, to that end, of full awareness of the role of language as a deformer of thought, inasmuch as, unfortunately, it plays that role as well, besides fulfilling a basic positive function in thought, in communicating the content of thought and in fixing its results. Making the scientific picture of the conjectural microcosm — inaccessible to sensory perception — independent of intuition, applicable only to macroscopic objects accessible to such, whether direct or indirect, perception. Replacing, wherever possible, the tentative, qualitative analysis of phenomena with a quantitative analysis — introducing wherever possible acts of measurement and calculation. Not chasing everywhere after regularities of a strictly deterministic nature, but — as broadly as possible — developing research of a stochastic nature, developing statistical methods and applying the achievements of the mathematical theory of probability. How many tasks arise here for institutions charged with fostering science! How much assistance can be afforded, having once understood the actual dynamic of changes in scientific methods and of the needs associated with these changes, if only by developing a solicitude for the continuing education of scientific workers (e.g., of those proficient in applying statistical methods). Among questions in the general methodology of the sciences belongs also no doubt the problem, oft-stated by ideologists of dialectical materialism, of transferring all scientific research onto the track of Marxist methodology (which requires undoubtedly in each instance a penetrating and unprejudiced interpretation).

An intermediate position between the general and special methodologies of the sciences is held by the problem of classification of the sciences. [13] For the latter concerns on one hand all the sciences as elements of a whole that encompasses them, while on the other hand what is involved are the features that differentiate certain sciences and groups of sciences from others. The task of science classification is a very ambiguous one. Sometimes cultivated by this name is something that is not classification at all, with its conditions of completeness and disjunction, but rather some sort of typology of the sciences or loose rosters of their varieties. In addition, what is subjected to classification or pseudo-classification is not always the same: sometimes these are sciences understood statically as systems of propostions, e.g., as that which constitutes the contents of books, and sometimes — sciences understood dynamically as complexes of operations comprising research specialties. One or another system arises as a result of the selected criterion of division, and the criteria are various depending on the purpose that the given division is meant to serve. One obtains quite different divisions if he differentiates the sciences for the most suitable disposition of printed matter in a library and for the preparation of the most operative catalog of the latter, and different divisions if he intends, e.g., to plan the choice and order for teaching of various sciences in a secondary school. We obtain a different result again by attempting, in the division of sciences, to reconstruct the natural, as it were, variety of processes whose regularities the various sciences are supposed to study. A different result again, when the constructor of the classification system is intent on building the entire system on a basis of piled-up dichotomies, which guarantee an authentic classification of the system (all-inclusiveness and non-overlapping of differentiated domains), but this virtue is purchased with the flaw of intolerable artificiality. And the opposite of such a speculative system, again, will be a system of the sciences corresponding to their actual professional division, a division by historically established occupations, or finally a projected optimal professional division of investigative and pedagogical disciplines.

Which, then, among the many possible ways of grouping the sciences promise the greatest utility in the rational division of social resources for science?[14] That is a cardinal problem. It is, however, impossible to consider it without prior reflection upon guidelines for dispersals of this kind. The objective would seem to be maximalization of the effectiveness of scientific research. But what constitutes effectiveness in this research? Most obvious here is the relation to the purpose of the scientific research as such; of course, it is the more effective, the richer and the more momentous for the blossoming of science are the cognitive results that it achieves or whose achievement it facilitates or makes possible. A second matter is how much the results of the given research contribute, by way of instruction in them, to making people more rational — a process that yields tremendous benefits in everything that people do, hence benefits as well from the standpoint of the third aspect of effectiveness, which involves productivity in the economic sense, in the sense of contributing to growth in production and improvement in quality of utilitarian goods. It is also important that all this happen at the least cost, in saying which we have in mind losses in the most varied resources used in scientific research.

And if we know what the effectiveness of scientific investigations consists in, then the question at once comes up: how can a society engaged in fostering science contribute to the increase of its effectiveness? Since all the forms of assistance practiced in this realm are surely known and since it may be assumed that they exhaust all the main possibilities, it will suffice to recall them briefly. Research institutions are established. The research institutions are authorized to hire personnel. Competent persons are hired, having earlier been appropriately trained. Laboratories are supplied with materials, with apparatus of all kinds and with every sort of ancillary object such as housing, reagents, instruments, etc. Funds are appropriated for these purposes. People are granted privileges, e.g., freedom of utterance in speech, writing and print. Technical possibilities are created for publication of the results. Institutions and operations are organized, set in motion and conducted which facilitate scientific work: libraries, information and documentation centers, etc.

Having precisely such intentions in mind, it is difficult not to take an interest in the mutual differences among the sciences on which must depend the rational differences in intensity and in ways of purposively influencing the effectiveness of research. In this respect, reflection and experience have already led to certain conclusions. Namely, we may divide investigations into those whose conduct requires basically, as ancillary objects, only paper, blackboard, writing instruments, manuscripts, printed matter and the like, and into those that in addition require other apparatus, e.g., experimental fields, chemicals, microscopes, measuring instruments, machinery, etc. Investigations of the first kind are incomparably cheaper, the second kind are — generally — incomparably more expensive. The foregoing division overlaps with another in the form of a quadripartite gradation. We have, first of all, directly usable studies, a distinct example of which is furnished by construction design. The type of corresponding questions can be formulated as follows: what operations are necessary or sufficient in order that thus-and-such may come into being?[15] Next come basic studies of two kinds — basic because they seek after relations, a grasp of which provides a basis for predictions as well as for plans of possible interventions leading to the intended product. These are directed basic studies, and free basic studies. Directed studies are those whose subject matter is made an object of scrutiny because the sought-after relation is regarded as being directly useful to the method of constructing a given desired product (e.g., the results of introducing a certain substance into animals' bloodstreams are studied, in order to use the relation that holds here as a foundation for the concept of a new drug). Designated as free basic studies, on the other hand, are basic studies whose content in only a general way promises such utility. An example is all chemical analyses of compounds of unknown composition that are conducted without plans for stated industrial applications. They do, however, have the character of basic studies on account of the general possibility of applying their achievements in the chemical industry.[16]

Finally, there are free studies not possessed of a basic nature. Their utility to the construction of products aside from a cognitive effect cannot basically be questioned, since the history of the sciences supplies numerous examples of the utility, not anticipated beforehand, of studies vastly remote, in respect of subject, from the field in which they later proved useful. Nevertheless, the basic value of such studies is their purely cognitive function, as well as their functions as a means of forming people's faculties and rationalizing their approach to everything[17] and as a cultivated way of filling the time left over after that devoted to work of a clearly economic nature.

Finally, a third division that overlaps with the previous two is the differentiation of scientific research of an ideological character and of scientific research that is divorced from complications of that kind. The circles that govern societies usually find difficulties, for example, in granting full privileges to social scientists in the public expression of controversial theses. Likewise subject to restrictions on verbal freedom are persons possessing expert knowledge of so-called national secrets, which sometimes belong in the world of technology. Hence the problem of a rational attitude toward these difficulties is a vital one.

Our review of problems in the science of science is running to a close. In conclusion let us be permitted to devote a couple of words to philosophy, for in a certain construction of that term a considerable part at least of the foregoing discussion undoubtedly belongs in that realm. Whenever we speak of philosophy, however, it proves useful to remind ourselves of the ambiguity of that term, which makes for difficulty in communication. By philosophy is understood fairly universally the complex of inquiries whose history the authors of works entitled 'history of philosophy' attempt to portray.

The objects of interpretations, critical analyses and genetic explanations here are chiefly, it appears, systems of views on the whole of reality and on reality as such, on the essence, object and limits of cognition, on the hierarchy of values that mark out the directions and principles of the most sensible, happiest and most respectable life. With the course of time, as the forms of higher education have taken shape, there has developed an interpretation—derivative of the first one—of philosophy as a certain set of subjects taught at universities. It finds reflection in lists of lectures at various universities, as well as in the catalogs of various libraries, where under the rubric of philosophy the patron will find metaphysics, theory of knowledge, logic, ethics and esthetics; and not so long ago psychology too was placed in this set. Competing with the above second understading of philosophy is an interpretation still closer to the first one as the summary, so to speak, of the results of the sciences at a given stage of development, and involved also is a summary integrated into a harmonius whole by the aid of connecting, as it were, hypotheses. In Marxist circles there is a dominant conception of philosophy as a system of laws governing the development of forms of existence and as a system of principles for cognition of these laws. To broader circles of thinking people, philosophy is a view upon the world and upon life, with the emphasis on life: hence, a kind of worldly wisdom that takes into consideration all the matters toward which it ought to take an attitude in order to be worthy of so estimable a name.

In addition to the mentioned interpretations of the term under consideration, there is still another, somewhat different construction. In this construction, philosophy is epistemology, the theory of human knowledge. Instead of elucidating this variant sense of the term 'philosophy' by means of isolated terms, let us make use of a perhaps somewhat too rough-hewn analogy. Philosophy is to human knowledge as economics is to human husbandry: it is the study of the essence and forms of its object of contemplation, the critique of its efficiency, a system of counsels serving its perfection. Can philosophy so understood be identified with epistemology? The question is an entirely legitimate one and must be attended to, considering the etymology of the term. *Epistemé* is, after all, the Greek for 'knowledge'. An interpretation of the term, however, has gained currency that is narrow compared with the term's genesis and composition. What is called epistemology is a fairly clearly isolated part of epistemology. What is meant is the tendencies of formation and functioning of the sciences as historico-social wholes, seen against the prospects for influencing their destinies by socio-organizational means. The interests of epistemologists have a historiosophical, a sociological, a socio-technical nature, and in fine are acquiring to ever greater degree a frankly economic character in face of the rational connection, through mutual dependence, between scientific investigations and undertakings of an economic nature. Consequently, on the grounds of epistemology, philosophy meets with social practice in an indispensable and jointly creative collaboration.

NOTES

[1] Cf. Ignacy Malecki, 'Z zagadnień metodologicznych nauki' (Some Methodological Questions in Science), *Nowe Drogi*, No. 3, 1964; also an article printed first in Volume XX of *Nauka Polska*, published by Kasa imienia Mianowskiego in 1935, by Maria Ossowska and Stanisław Ossowski, *Nauka o nauce* (*The Science of Science*), in the present volume, p. 82; it is full of germane ideas although with regard to terminological recommendations it deviates from what is proposed in our discussion.

[2] A possible source for recalling various problems in the science of science is the extracts edited by G. Picon as *Panorama des idées contemporaines* (Polish edition, *Panorama myśli współczesnej*, 1957, Paris, Libella).

[3] Cf. also J. D. Bernal, *Science in History*, London, 1956.

[4] Jean Piaget edits a series, published in Paris by Presses Universitaires de France, entitled *Études d'épistémologie génétique*, now embracing nearly a score of items and summarizing work done at Piaget's centre of experimental psychology in Geneva. For example volume XVII: Gréco, Inhelder, Matalon, Piaget, *La formation des raisonnements récurrentiels*, 1963; A. Michotte, *The Perception of Causality*, London, 1963; J. Pieter, *Praca naukowa* (*Intellectual Work*), Katowice, 1957 (rather questions of methodology in the so-called technology of intellectual work, seen through the eyes of a psychologist and pedagogue).

[5] For example see part three ('The Sociology of Knowledge and Mass Communication') and part four ('Studies in the Sociology of Science') of R. K. Merton's, *Social Theory and Social Structure*, 3rd ed., Glencoe, Illinois, 1959.

[6] Cf. S. Kamiński, *Pojęcie nauki i klasyfikacji nauk* (*The Concept of Science and classification of Sciences*), Lublin, 1961, p. 15. A rich and instructive catalogue of problems in the science of science. On cannot help noting, however, that some of Father Kamiński's statements on theology — e.g. p. 116, lines 13 and 35 — are flagrantly at odds with the character, in general sober from a lay point of view, of the book as a whole.

[7] "To M. Daumas... problems of the social determination of science and of its social consequences are not to be considered as belonging directly in the province of science historians... A different position was defended by J. R. Ravetz. In his judgment, science history was formerly treated as a history of scientific discoveries, whereas today science is treated as a social function, and precisely for this reason the history of science ought to be regarded as the history of people's cognitive activity, socially conditioned and conditioning subsequent social evolution", B. Suchodolski, 'Sympozjum w Jabłonnie poświęcone ogólnym zagadnieniom historii nauki i techniki' ('Jabłonna Symposium on General Questions in the History of Science and Technology'), *Kwartalnik Historii Nauki i Techniki*, V. IX, 1964, No. 1, p. 145.

[8] "...directors of establishments expect from... scientists in research laboratories... strict conformity... to... the firm's purposes. They say that there is no sense in financing the 'idle curiosity' of the scientists. This sort of approach... makes it impossible to take up new problems which can lead to results that will revolutionize production and multiply the company's profits. Confirmation for this thesis are the experiences of two leading American concerns — General Electric and Bell Laboratories. These concerns allow their scientists to choose for themselves, to some extent, the subjects of their own research. Thanks to this, C. Shannon, employed at Bell Laboratories, was able to occupy himself with so abstract a field as communication theory, which as a result... became a 'gold mine of highly practical applications'. It was pretty well the same way with the achievement of synthetic fiber at the Du Pont Company. Wallace Hume Carothers came to Du Pont from Harvard University, where he had studied molecular structures. Nylon was the side product of Carothers' actual, seemingly impractical scientific interests." A. Wilk, 'Uwagi o organizacji badań naukowych w Stanach Zjednoczonych' ('Some Remarks on the Organization of Scientific Research in United States'), *Życie Szkoły Wyższej*, 1963, No. 12, p. 95.

[9] "Science history conceived of as the history of the evolution of the mutual relations obtaining among scientific disciplines... could... help contemporary planners in discerning evolutionary tendecies leading to the future". B. Suchodolski, 'Sympozjum w Jabłonnie...', *op. cit.*, p. 145.

[10] Cf., among others, T. Kotarbiński, 'Dążności rozkładowe postępu wiedzy' ('Disintegration Tendencies of the Evolution of Science'), *Przegląd Filozoficzny*, 1915, issue 1–2; reprinted in *Wybór pism* (*Selected Works*), Vol. II, 1958, p. 255.

[11] L. Armand, 'Réflexions sur la recherche', *La jaune et la rouge*, publication mensuelle, numéro spécial de 1963 *La recherche fondamentale et appliquée*, Paris, p. 133 ff.

[12] Proposals for a different usage than in the present article of the term 'methodology' (in connection with the terms metascience and *metodyka*) are contained in a valuable article by K. Ajdukiewicz, 'Metodologia i metanauka' ('Methodology and Metascience'), *Życie Nauki*, Vol. VI, No. 31–32, 1948, p. 4 ff; cf. T. Kotarbiński, 'Treść i zakres pojęcia metodologii ('The Meaning and Scope of the Concept of Methodology'), *Myśl Filozoficzna*, No. 4 (24), 1956, p. 97; T. Kotarbiński, 'O pojęciu metody', ('The concept of Method'), *Zeszyty Wydziału Filozoficznego Uniwersytetu Warszawskiego*, 1957, No. 1; T. Kotarbiński, 'O istocie i zadaniach metodologii ogólnej' ('On the Substance and Problems of the General Methodology'), *Przegląd Filozoficzny*, Vol. 41, issue 1, 1938, p. 68 ff.

[13] T. Kotarbiński, 'Z dziejów klasyfikacji nauk' ('From the History of the Classification of Sciences'), *Życie Nauki*, Vol. V, Nos. 3–4, 1950, pp. 232–52; reprinted in *Wybór pism* (*Selected works*), Vol. 2, Warszawa, 1958, p. 360; B. M. Kyedrov, *Klassifikatsiya nauk, I. Engels i yego predshestvyenniki*, Moscow, 1961; L. Apostel, 'Le problème formal des classifications empiriques', and C. Perelman, 'Réflexions philosophiques sur la classification' in the collective work, *La classification dans les sciences*, Gembloux, Belgique, Éditions J. Duculot and Co., 1963. The articles mentioned do not pertain especially to the classification of the sciences but consider questions in the general theory of rational classification, which have applicability as well to the classification of the sciences.

[14] Cf. H. Jabłoński, 'Kryteria wyboru kierunków badawczych' ('Criteria in Selecting Directions of Research'), *Nowe Drogi*, 1964, No. 2, p. 46.

[15] Cf. Bernal, *op. cit.*

[16] *Ibid.*

[17] "It [sociology] is... a theoretical knowledge that furnishes people who are interested in the life of society with an orientation in complicated problems and helps them attain what C. Wright Mills called a 'definition of one's own situation'. This humanistic practical function of sociology does not lead itself to calculation in terms of concrete, measurable, practical benefits, but is indispensable if one wishes to realize a society of people consciously shaping their own future... Some sociologists... consider precisely this function of this science to be basic". J. J. Wiatr, 'Współczesna socjologia polska' ('Contemporary Polish Sociology'), *Życie Szkoły Wyższej*, 1963, No. 12, p. 68.

Bohdan Walentynowicz

DECISION THEORY, SYSTEMS THEORY AND APPLIED RESEARCH*

Translated by Christopher Kasparek

Purpose and scope of the paper

In the present paper I will attempt to show that two recently emancipated disciplines, decision theory and systems theory, are very useful to scientific research and that the idea of using them to increase the effectiveness of such research deserves popularization.

The subject of the paper being thus formulated, in view of the limited diffusion of those two theories it seems proper to restrict the discussion to a statement of basic and general — perforce at times very broad — remarks and to eschew the ambition of making a personal contribution to a particular selected question.

I shall limit the scope of this paper moreover to a discussion primarily of the introductory phases of any scientific investigation, i.e., to the phases associated with formulation of the objective, and also primarily to investigations conducted in the so-called applied sciences.

In view of persisting terminological ambiguities, immediately following an introductory discussion of the possibility of creating metasciences of universal methodological applicability I will describe more closely what is to be understood by research in the applied sciences — that being the chief subject of this paper — in order

* First published as 'Teoria decyzji i teoria systemów a badania stosowane' in Polish journal *Zagadnienia Naukoznawstwa,* Vol. 1 (1965), Nos. 2–3.

next to sketch the fundamentals of decision theory and systems theory and finally to analyze their applicability to the kind of research under discussion.

The creation of metasciences of universal methodological applicability

It may generally be said of research methods that they take form and are perfected primarily within the framework of the various fields of knowledge. The transfer of a given method to another field often is out of the question, so very specific is the method to the field in which it arose. Nonetheless, one also encounters methods of considerably greater universality, i.e., ones applicable — and with great benefit — not only in the fields from which they sprang but also in other, often remote domains.

In addition, every so often there develop independent disciplines and theories that have a great and broad methodological applicability, as for example various fields of mathematics that have come into existence during the history of learning and that have always exerted great influence on the methods applied in a given period.

Recent years have witnessed development of many new fields of learning to which may be attributed precisely this kind of methodological importance. Among these may be numbered the following sciences: value theory, dependence theory, praxiology, operations research, game theory, cybernetics, management science, organization science, and also — the subjects of this paper — decision theory and systems theory.

The methodological utility of these sciences depends on the possibility of their application, hence on the possibility of establishing a correspondence between concepts occurring in these sciences and phenomena encountered in scientific research.

The above-mentioned sciences take their genesis in a process of generalization in which we remove details regarded from some standpoint as being less important and leave those that seem to us worth investigating.

We may grant to both decision theory and systems theory this degree of generality, while noting that the laws formulated by them are not trivial: that the emergence of these disciplines benefits scientific endeavours by highlighting certain special aspects of the problems under consideration.

Uyomov [1] has interpreted the foregoing features of these disciplines, which endow them with generality, as being features associated with their special character and as making of them a new category of sciences.

Characterization of the practical sciences and of their methodologies

If I say that I propose to limit this discussion to research conducted within the practical — known also as the applied — sciences, then it is incumbent on me to indicate, if only briefly, the features distinguishing these sciences from the sciences termed pure, basic or theoretical. It will suffice for purposes of subsequent discussion if I emphasize that 'pure' sciences study that which exists or which, according to prior data, could exist. The practical sciences, by contrast, study that which for some reason someone might wish to be made to exist through means elaborated for that purpose by those sciences. In other words, I have in mind sciences whose purposes touch not only a desire to increase the store of knowledge without thought to its direct use but, on the contrary, that touch knowledge that makes possible the resolution of various concrete difficulties generally encountered by someone outside them. Research conducted within these sciences is most commonly termed applied (technical) research.

The sciences thus defined include, of the traditional natural sciences: the medical, agricultural and technological sciences; and of the traditional social sciences: economics, business, law and education. The distinction between the above sciences and the so-called pure, theoretical or basic sciences, as already indicated in the foregoing definitions, may be further supplemented by the following important observations.

The purposes of the 'pure' sciences involve what we call 'determining the truth', whereas those of the practical sciences go beyond the scope of the pure sciences and involve cognition that makes possible the realization of some desired thing.

Certain difficulties may arise in distinguishing these sciences from analogous practical activities, e.g., the medical sciences from medical practice, the technological sciences from technological practice, etc. They are not set apart by their purpose, for they may serve the same ends. Rather they are set apart by the concreteness of the objects, situations and purposes considered. For whereas the practical, e.g., the technological, sciences consider the entire class of objects in terms of certain generalized properties, the entire class of situations involving these objects and the entire class of possible purposes, in practical activity we are concerned exclusively with a definite object existing in a given situation or with a set of such objects and situations, possessed of a probability distribution of attributes known to a degree sufficient for decision making, and with a given set of purposes characterized by mutually appropriate evaluations of their validity.

There are several reasons why analysis of the utility to scientific research of the theories under discussion may profitably be concentrated on applied research. The methodology of the practical sciences has been very inadequately worked out, due perhaps, among other things, to their specializing tendency and interest in unique solutions. But the importance of methodological clarity in fields in which problems are often indistinctly drawn is enormous, and the social results of increasing the effectiveness of applied research — which directly facilitates the avoidance of many errors in practical action — are far-reaching. What is more, despite the widespread view that 'pure' research involves a higher degree of methodological elaboration than does applied research, it is in fact applied research — to which a concrete, e.g., an economic, purpose may often be ascribed — that better lends itself to methodological treatment than does basic research.

The real reason, however, for formulating the subject of this paper in just this way is different. It is that the benefits of applying decision theory and systems theory are — as it seems — at their most direct and indisputable in the above-defined sciences.

The decisional problem model and decision theory

Any problem, and hence any situation involving research problems, can be represented symbolically as:

$$U = f(A_i, S_j) \quad i = 1, 2, 3, \ldots, n, \quad j = 1, 2, 3, \ldots, m$$

where U is the *value of the utility* of that which is to be optimized (e.g., states of a system, of an operation); A_i are aspects of the situation that are subject to steering, or *decisional variables* characterizing actions taken, conclusion drawn, etc.; and S_j are aspects of the situation that are not subject to control, or *parameters* characterizing the context of the problem.

Solution of the problem consists in finding such values of the decisional variables A_i, expressed in a function of variables S_j as maximize (or minimize) the value of U.

The decisional problem is often represented in matrix notation, wherein the various rows correspond to alternative actions (operations, procedures, views) described by decisional variables, and the various columns to situations (conditions of application, states of reality) described by parameters characterizing the context of the problem. Establishment of a criterion of optimality permits selection of that row in the matrix for which utilities U of the results R_{ij} in cases of various situations show a maximal value.

	Alternative 'situations' S_j (sets of uncontrolled parameters)
Alternative 'actions' A_j (sets of decisional variables)	$U(R_{ij}) = U(A_i, S_j)$

Analyzing this more precisely, we can list, e.g. following Ackoff [2] but also in accordance with other authors, [3] the factors whose existence is necessary in order that we may say that we are dealing with a certain problem:

(1) *A person facing the problem*, experiencing doubts concerning the correctness of a choice, and possessing the freedom — as well as the obligation — to make a decision.

(2) *A result* R_{ij} constituting for the decision maker, in the case when it is a desirable result, a goal to be attained dichotomously (as 0 and non-0) or by gradations and being, in general, an aggregate result of results (and thus of goals) that are partial, simultaneous, not infrequently mutually contradictory, either qualitative — subject to dichotomous limitations — or quantitative and thus measured by certain evaluative scales and subsequently reduced to a single scale.

(3) Initially proposed *actions* A_i (procedures, hypotheses), at least two, objectively of unequal effectiveness and distinctly different, available to the decision maker and steerable, and corresponding at times — in the case when they constitute a basis for selecting a procedure — to strategies of procedure.

(4) The *context* of the problem, or conditions S_j brought about by various factors influencing the result and not subject to steering, among which may be distinguished primarily circumstances characterizing 'the state of Nature' as well as other persons exercising freedom of decision and interacting either collaboratively or conflictingly (e.g., competitively).

The matter of supreme importance in the decisional problem so formulated is *the criterion of optimal choice*, i.e., that which is to be understood by the concept of the 'best' or 'optimal' solution. No general definition of what is called 'best' in variouss contexts has up to now been given in science, and there is little likelihood of one ever being formulated. Nevertheless, the statistical theory of decisions in a normative elaboration may note certain achievements in approximating to a precise concept of the 'optimality of the solution to a problem'. [4]

The person making a decision is not always at the same time the person carrying out studies whose end is to solve the problem. In the general case these are different persons. For the person who is faced with the problem and who makes the decision 'commissions' the researcher's solution. Although he does not personally solve the problem, he is nevertheless named as an important factor in the decisional problem, since in the absence of an objective value criterion the problem is solved for this person, that is to say, for utility understood in the way that he understands it, for preferences formulated by him. It goes without saying how important it therefore is that the executive decision maker and 'research commissioner' should not be a dilettante but an accomplished specialist in the field that the problem pertains to.

Criteria of optimal selection are associated with another matter of capital importance in decisional problems: with the reliability of the information possessed and with the possible necessity of *predicting*. For these criteria depend on whether the information involved in the context of the decisional problem, and also in the possible procedures, is given in a determined manner or merely probabilistically, and also on what kind of information exists on the results of applying various procedures in various contexts, with these results being expressed in the form of appropriate relations.

Depending on what kind of information the decision maker is dealing with, he may be faced with the necessity of making decisions in conditions of: (a) certainty, (b) risk or (c) uncertainty.

In conditions of *certainty* all the information has a determined character, there are no conceptual difficulties in evaluating utility, and the question of the decision may be regarded as trivial, provided only that one knows how to assign degrees for the attainment of the individual objectives, to ascribe certain utilities to them, and so to weigh them as to obtain a resultant utility.

In conditions of *risk* it is customary to subject to optimization the expected value of the result of applying each of the possible procedures (the 'mathematical expectation'), and hence in the case of the possible occurrence in a given procedure of the alternative results (and only of the results) R_1, R_2, R_3, ..., with probabilities estimated as being respectively p_{11}, p_{12}, p_{13}, etc. $(p_{11}+p_{12}+p_{13}+ \dots = 1)$, the magnitude

$$E(R_1) = p_{11} R_{11} + p_{12} R_{12} + p_{13} R_{13} + \dots = \sum_{j=1}^{m} p_{ij} R_{ij}.$$

The above criterion, correct intuitively, can also be justified, even though various possible reservations as to its correctness have so far not been incontrovertibly removed. The sense of this criterion depends moreover on whether the objectives in the problem under consideration are *qualitative* (expressed as zero or one) or *quantitative* (expressed along a certain scale) or, finally, *mixed*, and on whether consideration need be given to interventions of other participants in the action through their collaboration or antagonism bearing on the outcome of the decision.

In conditions of *uncertainty*, estimated values for the probabilities p_{ij} of occurrence of results R_{ij} are unknown. The decision maker is not helpless, however, since there exist various proposed criteria of procedure in this situation (e.g., of the 'minimax' or 'maximin' type). To be sure, their validity so far in science is till more disputed than that of using the criterion of expected value in conditions of risk.

These criteria have been analyzed lucidly by Szaniawski, [5] who has also put forward an original proposal for a new criterion of this kind.

There is still another possibility here, however. It is sometimes possible — using certain concepts of game theory — to reduce a case of uncertainty to one of risk, which makes it possible to apply the criterion of expected value. For if the situation would have to be recognized as being unrealistic in which all the information is in fact given with complete certainty, so the situation in which one succeeded in formulating the problem but possessed no information concerning estimates for probability of occurrence of certain results and not of other results, should likewise be regarded as being unrealistic. The information that sufficed for the formulation of the problem (for the designation of pertinent procedures, of possible situations of application, and so forth) frequently suffices for estimates for the values of probabilities of occurrence of given results or — which is also at times sufficient to reduce the situation to one of a choice in risk conditions — for setting certain limits to the value of these probabilities.

The concept of system, and systems theory

Systems theory is closely linked with decision theory. The problems with whose solution decision theory concerns itself do not, after all, exist in a vacuum, and possible states of the world as well as possible procedures both ought to be studied and described. Such description is facilitated by the field whose object of interests comprises every possible system in which some realization takes place, and which in turn makes use of decision theory in shaping systems and their working states in one way and not in another.

It is, however, no easy matter to give a generally binding definition of this new, scarcely a score of years old, scientific discipline known as *systems theory* or *systems science*, and in somewhat different contexts, as *systems technology* or *systems analysis*. Many pages have been devoted to the attempt to define systems theory, yet there remain great differences among definitions given by various authors. It is interesting that the idea of systems theory probably originated with biologists, if in point of fact with biologists known for their cybernetic interests and achievements (Ashby, Bertalanffy); immediately thereafter the most interest was shown in this field by technologists and economists.

It would seem adequate initially to content ourselves with an attempt at defining the concept of 'system', in order that the subject matter, scope and treatment of 'systems theory' may become intuitively of sufficient clarity.

In everyday speech the term 'system' is used in a very broad sense. One speaks of a philosophical system, a number system, a communications system, a regulatory system, an educational system, a political system, a defense system. [6]

Leaving aside the sense related to the concept of manner, a system may be said to be an organic or organized whole composed of independent parts, with the proviso that systems theory is interested only in systems composed of parts carrying out a certain operation and in systems that are at least to a degree steerable, unlike for example the solar system. The parts of the system may be spoken of as its members, organs, components or elements, and in the general case these terms may not be synonymous.

According to one author's definition, a system is that which accomplishes an operational process, i.e., that in which something is operated so as to produce something else. [7] According to another author's definition, a system is a set of objects and relations among objects and their proprieties. [8] According to yet another difinition, systems are engineering artefacts most easy analyzed, described or designed as assembly of simpler part.[9] These definitions are not entirely equivalent in meaning; considerably less controversial are formalized notations of a system given in the literature, but which we must forgo citing in this paper.

There are infinitely many possibilities for the mental isolation systems from the surrounding world of things and concepts, and because of this it is necessary in each instance to formulate clearly in what way the system described is treated at the given moment as an object of study.

In view of the arbitrariness of system isolation, it is obvious that any system encompasses various lesser systems that are *subsystems* in respect to the system in question, and that at the same time it belongs to other greater systems constituting *supersystems* in respect to that system. In this way every system comprises part of some supersystem, along with other systems of that supersystem. Thus the other systems constitute, with reference to the given system, an *environment* that influences the operation of that system.

If we wish to *analyze* a system, we isolate the component subsystems within it; if we wish to make a *synthesis* (in accordance with certain 'commissions'), we create in a selected manner a supersystem of lesser systems.

Sometimes the question arises, why not expand the system under consideration to one degree or another by systems that constitute the environment or by parts of systems, if consideration is given to their influence on the system. Such an approach is also possible, since the formula selected is a matter of convention and one should follow convenience in reasoning.

Three categories of systems may be distinguished:[10]

(a) *natural* systems, of organic or physical origin, containing many components not steerable by man;

(b) *invented* systems, planned and worked out by man; the organic or physical material comprising them, and their forms, are modifiable by human intervention;

(c) *hybrid* systems, constituting combinations of natural and invented systems.

Within any system a given process is realized, and the relations among the several parts of a system depend on their cemmon participation in the system's flowpaths of information, energy, materials or objects. All these factors can occur in a system under various guises, which might be termed — in accordance with the information flows — the 'language' of the system.

Within the system one can generally isolate an *input* and an *output* or various inputs and various outputs of an informational, energetic, material, objective, etc., nature: while the system's action, manifested in the system's transformation of various inputs into outputs, may be either determined or stochastic. A description of the system requires an indication of its current state. The set of possible states of the system forms a phase space. The course of the process underway in the system depends on its current state, and frequently on previous states as well.

Such a formulation of the concept of a system is a universal one: it is equally suitable to various working processes, physical and biological objects in which processes occur, and to conceptual entities made up of variables.

In order to describe a given system it is necessary to identify the character of the outputs, the character of the inputs and the set of possible states for the system, as well as to formulate a *model* expressing the relations obtaining among the several inputs and outputs and reconstructing the course of changes in the system's states over time.

A more minute description will require a progressively more exact knowledge of quantities occurring in the system and of their mutual relations as well as of the functions characterizing the processes. Systems theory concerns itself precisely with the development of these methods and techniques of procedure. Their usefulness is the greater the more complex is the system under consideration; and in certain cases the investigation of complex systems would be altogether impossible without recourse to procedures elaborated expressly for that purpose. This approach is such a general one that techniques of analysis worked out for a given system, e.g., a biological one, can be carried over to another system, e.g., an industrial one, provided only that the models representing the respective systems are similar. Deserving of particular attention in analysis is the identification of the processes and relations from the standpoint of whether they are determined or stochastic in character.

An enormous role in the investigation is played by the degree of the system's complexity. The best known scheme for classifying systems from this standpoint is Boulding's [11] given below.

TABLE 1

Hierarchy of systems

Level	Typical representative	Characteristic properties	Example
1.	Structure	Stability	Earth's geography, organism's anatomy of cell
2.	Machine	Predetermined motions	Steam engine, solar system
3.	Homeostat	Regulation	Thermostat
4.	Living cell	Open system of continuous material and energy flow	Flame, river, cell
5.	Plant	The processing of (non-psychical materials	Any plant
6.	Animal	Self-preservation, specialized senses, the capacity to process information into ideas (distinct from the information)	Any animal
7.	Man	Self-consciousness, language, symbolization	Any human
8.	Social organizations	'Roles' and communication, history, knowledge, music, art	Societies
9.	Transcendental systems	Unknown	Unknown

Our knowledge of systems at a given level is the smaller the higher the *level of complexity* of the system. According to Boulding, most of what we presently know about systems pertains to the first, second and third levels. We are only beginning to develop a knowledge — in a systems-theoretical sense — of systems at the fourth level of complexity. Nonetheless, the systems-theoretical formulation of complex systems greatly facilitates their investigation.

The character of a system and the manner of its investigation are greatly influenced by whether or not the system involves people. Many of the problems currently recognized by science as being urgent relate to systems composed of machines and men. These systems show a capacity for being at least partially *steered*, i.e., they permit some freedom in choice of purposes and of ways to attain these, given existing limitations, thus making these systems organized systems, or *organizations*.

A fundamental matter in characterizing systems of the organization type is the identification of information flowpaths and of decision-making loci, of the ways in which tasks, responsibilities and competencies are divided up.

A crucial matter in systems theory is that of *evaluating* various systems and system states with reference to the system's objectives. This evaluation, carried out on the basis of certain established criteria, makes it possible to take decisions involving changes of system states from existing ones to more desirable ones.

The elaboration of procedures and criteria for evaluating systems and system states constitutes one of the larger divisions of systems theory. The usual approach is to formulate so-called *evaluative standards*, i.e., categories of generally binding evaluative criteria which subsume evaluative aspects binding in the given case; that is, translated as it were into the language of the system.

Among these standards, variously listed by different authors, the ones that occur oftenest — e.g., in the case of systems that produce certain kinds of technical equipment — are efficiency (the degree of performance of a task, formulated in different ways in accordance with the character of the system); cost effectiveness (with respect to initial outlay, with respect to exploitation costs — in absolute terms or with reference to another quantity, as unit costs or indicators); reliability; serviceability; availability; flexibility (e.g., ease of modification); simplicity; safety (e.g., for the environment); aesthetic qualities; degree of advancement; etc.

The foregoing list of viewpoints considered in the evaluative process, together with what was said earlier, indicate the nature of the new science called systems theory or, in reference to systems involving technological resources, technology or systems engineering. Its novelty resides in the fact that, while resting on many other already existing sciences and methods, it treats emerging problems as wholes.

A systems treatment of a given problem consists in a process whereby — by contrast with a process of optimizing individual components of a system or of considering sequentially individual viewpoints in the conviction that the system thus formed will be an optimal one — the system as a whole is optimized, regardless of whether or not this is optimal if one considers only one component or only one standpoint.

It follows from this that complex problems cannot be studied effectively — by way of operations research, as an applied decision theory, and by means of systems theory — by single investigators. The complexity of the problems, and hence the broad array of capacities essential to the solution of these problems, generally require the collaboration, as members of research teams, of mathematicians, logicians, statisticians, lawyers, economists, sociologists, biologists, physicists, technical specialists and others.

The utility of decision theory and systems theory to scientific research

The nature of research conducted by the practical sciences, and the foundations of decision theory and systems theory sketched out above, make clear the utility of these theories to the research under consideration.

The direct connection between applied research and the procedures developed in the above theories is immediately evident if we bring together the respective definitions once more and say that:

(1) applied research involves a search for optimal solutions

to problems and the planning — for the system under consideration — of a state that might be regarded as the closest possible to the ideal state recognized as desirable;

(2) decision theory occupies itself with seeking optimal solutions to problems, formulating them in decisional models, and elaborates operational procedures of optimization;

(3) systems theory is concerned with the description, analysis and synthesis of all kinds of systems and works out procedures for designating their optimal states and for their planning, equipping with various technical resources, etc.

In order to make use of these theories in applied research, the investigated problem must be translated into the terms of a decisional problem and of a 'systems' problem, i.e., the problem in question must be formulated in the language of decision theory and systems theory.

Inasmuch as decision theory and systems theory are closely linked to one another — and this to such a degree that it is impossible to begin speaking of one without speaking of the other (systems theory makes use of decision theory, and every decision basically involves some system) — there is no need to make two separate formulations. Even though the degree of generality of these theories is different, one may use a language common to both theories.

Expressing the problem in the language of these theories constitutes an excellent identification of the problem, which is the first methodological benefit of applying decision theory and systems theory in applied scientific research. One verifies whether he is really studying the right problem in order, should it prove to be not so, to reformulate the problem in a way that is more fitting or easier to solve.

After formulating the problem one should apply the procedures of solution developed by these theories. These are scientific procedures, verified in respect of internal consistency, economical in use. The benefits from their application are considerable. They render explicit the inadequacies in the existing information and its character (determined or probabilistic), the assumptions made, possible sources of error, the relations obtaining among the quantities comprising the model, the correctness of the limitations considered, the criteria of evaluation adopted, etc.

There exist problems — and there are more and more problems of this kind — in connection with which resort to the conceptual apparatus and procedures elaborated by systems theory is altogether indispensable if one is to attempt their solution. These are problems involving complex systems, e.g., the systems studied by astronautics. Such problems are studied more and more often since, on one hand, the world surrounding us is becoming increasingly complex due to ever richer and more sophisticated technical equipment and due to its increasing interconnectedness, and on the other hand in our desire to master this world — if only intellectually — we make resort to the scientific solution of problems involving complex systems and study more and more things as great wholes. In the book edited by Eckman, [12] the author of the foreword, S. Ramo, presents the matter most dramatically, writing that we are witnesses to a race between the constantly growing complexity of systems and the development of theories and techniques permitting their study.

We could cite many examples of problems in which we draw clear benefits from applying procedures elaborated by decision theory. These include situations, frequent in scientific technical studies, when on the basis of experimental data S_j we are to give a plan A_i of procedure for practitioners so that they may secure stated results. The plan A_i may involve a way of treating a patient's illness by a physician, a formula for calculating some mechanical element, a function approximating the curve of some characteristic, e.g., of chemical properties, and so on.

A subtler reasoning process leading more quickly and reliably to a conclusion correct by accepted criteria is afforded us precisely by decision theory, especially by statistical decision theory, in which it is assumed that the information on particular S_j's is incomplete.

Decision theory and systems theory as sufficiently general theories make it possible to use solution techniques and strategies applicable to diverse, sometimes very disparate fields, provided only that the models of problems considered by these fields have the same form. This may be illustrated by one of the best known, models in operations research, the queuing model, which can be used to calculate the optimal size of staff necessary to serve clients, the optimal extent of port facilities, the number of airport runways etc.

The methodological usefulness of decision theory and systems theory is apparent in scientific research even when ultimately, for one reason or another, we must give up the attempt to apply to research a decisional model or some technique of calculation worked out within these theories. The compelling rigours of disciplined thinking flowing out of these sciences increase the efficiency of research even when ultimately they cannot be applied.

In regard to the most difficult matter in problem solving, the criterion of optimality, decision theory concerns itself with analyzing the rationality of criteria posited by goal setters, but these criteria basically constitute the subject of the general theory of value, a discipline that is in the first stage of its development and that shows no clear developmental prospects, and in the final analysis these criteria cannot but have an ethical character.

As far as concerns the other crucial matter in problem solving, i.e., the necessity of predicting, the techniques of seeking solutions described in the literature of both decision theory and systems theory consider this to a large extent. One author estimates that the number of problems solved with the aid of numerical techniques taking into account the stochasticity of the processes under consideration will rise due to the development of systems analysis and operations research [13] as shown in the following table:

		determined (%)	stochastic (%)
hitherto	analytically	98	1
	numerically	1	0
in future	analytically	55	25
	numerically	15	5

And in fact we do observe the development of applications of stochastic processes of probability and of numerical computations. The complexity of the resultant calculations is compensated for by the development of numerical techniques that make possible their execution by electronic machines. Many pages in books on decision theory and systems theory are devoted to discussion of these fairly complex computational procedures.

To the view that holds that the application of procedures elaborated by decision theory and systems theory is useful in applied research, the following reservations may be advanced, and to these in turn the folllowing arguments opposed:

(1) Decision-making in the course of problem solving requires principally the possession of information, and neither decision theory nor systems theory increases the stock of information.

Answer: It is true that these theories do not directly increase the stock of information and that, when applied to research within a given field, they answer only for logical and not for substantive truth. They do, however, show what information is indispensable or desired and supply procedures for evaluating the significance of attainable information from the standpoint of the investigator's work; moreover they instruct him how, by applying proper procedures, to make optimal use of the information resources in his possession without suffering information losses and how to use these effectively. All this cannot but increase the effectiveness of the pursued research.

(2) Problems as formulated by both the above theories become complex and the selection procedures — considering, among other things, the necessity of predicting, the stochasticity of processes, and the unreliability of much information (amounting to its lack) — cumbersome in application.

Answer: It is not the theories that complicate the problems, but the problems themselves that are complicated — often more so than would appear at first blush. It is the merit of these theories, among other things, that they reveal that fact. Surely it is more appropriate to use subtler modes of reasoning that consider the problem in a correctly formulated fashion and hence often treat it as complex, rather than to succumb to the illusion that it suffices to employ a more primitive reasoning process because the problem is thought not to be complex. It is not decision theory or systems theory that introduce stochasticity into the model of the problem being weighed. The stochasticity in many cases must be accepted as existing regardless of how troublesome it might be. In both the above theories, procedures are worked out that consider stochasticity, yet lead to a solution.

(3) Both theories advocate the formulation problems without the use of simplifying assumptions, in consequence of which the problems so formulated become either altogether insoluble or tremendously cumbersome in their solution. The advantage in a simpler formulation of a problem is that this assures its solubility.

Answer: Neither of the above theories can forbid the use of simplifications, if the latter constitute a condition for solubility. There is, however, a difference between a simplified formulation and the conscious adoption of necessary simplifications following an initial formulation without simplifications. In the second case the simplifications adopted are expressed explicitly and the error stemming from individual assumptions can be more readily calculated. From the methodological point of view a considerable superiority must attach to a method in which the repertoire of assumptions is explicitly given. This enables subsequent students to benefit not only from the substantive but also from the methodological achievements secured through the research.

Conclusion

The matter of the utility of decision theory and systems theory to applied research may be treated as a separate problem — a methodological problem — and weighed as such.

This problem may also be considered using procedures of decision theory and systems theory. In this case the method based on decision theory and systems theory will be one of the possible procedures in the decisional model. Solution of the model requires — as in any case — overcoming difficulties connected with the collection of indispensable information, and the establishment of the decision criterion.

The popularization of group studies in the face of the ever greater complexity of systems and of the resulting necessity of such research creates interesting methodological problems. Both decision theory and systems theory may prove enormously useful in solving problems in which the system under consideration is a group of researchers and their tasks, and the possible actions are various principles and forms of their collaboration among which one wishes to find a solution that is optimal in the light of accepted criteria.

One may imagine yet another problem, taking into consideration a system otherwise defined and regarding it again, perhaps, by a method based on decision theory and systems theory. Specifically, one may wonder whether the solution to a problem is to be sought by way of scientific studies or by way of procedures based on experience, intuition, or by way of other procedures.

It seems that decision theory and systems theory, in contributing to the more precise formulation of a problem, will manage to demonstrate that, despite the direct costs involved, scientific research methods based on these theories are less expensive and assure more rapid progress than ostensibly cost-free and operative non-scientific methods, including intuitive methods and arbitrary methods, as well as methods of the trial-and-error type.

NOTES

[1] A. I. Uyomov, 'Nekotorye tendentsii v razvitii estestvennykh nauk i printsipy ikh klassifikatsii', *Voprosy filosofii*, Vol. XV, No. 3, 1961, p. 67.

[2] R. L. Ackoff, *Scientific Method: Optimizing Applied Research Decisions*, Wiley, New York, 1962.

[3] R. M. Thrall and others (eds.), *Decision Processes*, New York, Wiley, 1954; H. Chernof and L. E. Moses, *Elementary Decision Theory*, Wiley, New York, 1959.

[4] Ackoff, *op. cit.*

[5] K. Szaniawski, 'Some Remarks Concerning the Criterion of Rational Decision Making', *Studia Logica*, 1960, No. 9, p. 221.

[6] D. P. Eckman (ed.), *Systems Research and Design*, Wiley, New York, 1961.

[7] D. O. Ellis and F. J. Ludwig, *Systems Philosophy*, Prentice-Hall, Englewood Cliffs, 1962.

[8] A. D. Hall, *A Methodology for Systems Engineering*, Van Nostrand, Princeton, 1962.

[9] W. Goasling, *The Design of Engineering Systems*, Heywood, London, 1962.

[10] Ellis and Ludwig, *op. cit.*

[11] K. E. Boulding, 'General Systems Theory: the Skeleton of Science', *Management Science*, April 1956, p. 197.

[12] Eckman, *op. cit.*

[13] H. H. Goode, 'Operation Research and Systems Engineering', in *Recent Advances in the Engineering Sciences*, McGraw-Hill, New York, 1958.

Jan Kaczmarek

ON PRINCIPLES OF SCIENCE POLICY*

Translated by Jerzy Łowiński

General characteristics of science policy

The term policy was used at one time to mean solely the art of achieving power in and ruling a state. With the passage of time however, the term became increasingly elastic and assumed various other meanings. Speaking most generally, the term at present denotes the principles of and skill in cultivating a certain field of human activity. All the present meanings of 'policy' continue to retain elements of authority and governement. Furthermore, the term is always connected with certain ideological categories and certain theses applicable to social, cultural and economic development.

Some time ago people began to use the expression 'science policy', meaning the steering of scientific activity broadly understood. The appearance of this term was due above all to the fact that science began to exert an increasingly strong influence on human life as a whole. Science increases the range and depth of human cognition, produces new world outlooks, enriches culture, develops economy, creates new social forms and contents.

Nearly all countries, despite different socio-political systems and different levels of civilization and cultural development,

* This article is an expanded version of a paper read at the plenary meeting of the Committee of the Science of Science of the Polish Academy of Sciences held on 1 July, 1971. First published in English in a Polish yearly, *Problems of the Science of Science*, 1971.

recognize the need to stimulate the mutual bonds between science and human activity, and do so in practice. They start out from the conviction, confirmed by experience, that methods of stimulation properly used, yield results which are from the social point of view better than if the links between science and life were left to be forged at random.

First of all, the view is propagated that through proper planning it is possible to improve and increase the effectiveness of all activity.

This view has emerged, above of all, in countries which base their activity on methods of planned development — that is, primarily in socialist countries.

Science policy — in the contemporary sense of the term — represents, therefore, a system of principles and of associated stimulating activity, whose aim is the shaping of such bonds between science and various manifestations of human activity as would permit the achievement of a desirable future situation and certain intended goals in the following spheres:

— non-material, which include the enriching of scientific knowledge and culture and the shaping of world outlooks;

— material, by which we mean all domains of economy (production and services).

This means that the characteristic feature of the theoretical foundations of science policy is that they have the nature of formulae determining the ways and methods of activity and transformation.

Optimum from the social point of view is a solution which permits or facilitates the simultaneous achievement of goals in both the non-material and the material sphere.

Comparatively the greatest development is observed as regards that part of science policy which is directed principally towards a proper mutual interaction of science and economy. [1]

It is, therefore, fully justified for science policy to take also an interest in technology which, together with science and economy, constitutes a specific triad of development in the material sphere, and has also a marked effect on the non-material sphere.

In the very nature of things, therefore, science policy applies not to science alone. The term science policy in its present and full meaning should be treated as conventional for the sphere of science and technology, and partly also for the sphere of economy.

This is connected with the differentiation of tasks arising from the cooperation of science, technology and economy into research work, developmental work and applications, where developmental work means construction, technological and investment designing.

Science policy so conceived embraces the following problems: (1) determination of directions and goals of development; (2) securing the achievement of goals and development in the directions determined; (3) exploitation of the effects of research and developmental work for the needs and use of the people.

Apart from the differentiated specificity of science, technology and economy, account must be also taken of the different specificity of activity in the material and non-material spheres.

The theoretical foundations of a science policy are created by the ideology of the socio-political system and a certain number of scientific disciplines. Worthy of special mention among those are:

— the history of the development of sciences and scientific-technical activity, as well as of cultural and economic development;

— the philosophy of science, with special reference to the essence of cognition;

— the methodology of scientific work (of cognition, creative work and communication);

— the sociology of scientific and technical milieus and groups;

— the psychology of scientific, technical and other staffs participating in research work and development work;

— the organization of scientific and technical work;

— the economy of science and technology, connected with the economic policy of a country.

In addition to the above disciplines, it may also be necessary to use other disciplines for the formulation of the principles of science policy, including political economy which enables the establishment of a closer bond between science policy and political ideology.

A considerable role in the above disciplines falls to the science of science, the development of which is comparatively recent. That is why there are views to the effect that the science of science constitutes the theoretical foundation of science policy. In the light, however, of a comparison of the scope of science policy with the generally recognized scope of the 'science of science' it appears that in its present stage the latter cannot be a sufficient theoretical base of science policy.

However, the science of science may be regarded as a domain subject to evolution, enabling it to meet all the needs of the scientific foundations of science policy.

Changes which seem to confirm this evolutional trend have been taking place in the science of science in the course of the last few decades. It may, therefore, become possible in time to differentiate between the theoretical foundations of scientific policy and the science of science understood not only as a collection or selection of the problems of numerous independently developing scientific disciplines, but also as a single, integrated scientific discipline capable of independent development.

Irrespective of the theoretical foundations on which the principles of science policy are formulated, their characteristic feature is that they help us to find optimum solutions in the situation of a given country. And this has an effect not only on the adoption of effective solutions but also on the very formulation of the principles of steering, which would place emphasis on strivings for the establishment of suitable, compromise proportions between phenomena and facts which either supplement or contradict one another. This is connected with a characteristic differentiation of certain principles of science policy depending on the size and the degree of development of a given country and its politico-social and economic system.

The principles of science policy presented below contain many theses which may be accepted and applied in different countries. In their entirety, however, they apply to Poland — that is, to an economically medium-developed socialist country with an average economic potential as compared with other countries.

The theses have been largely formulated by way of conclusions drawn from ideological premises and generalizations of practical experience, which means that they are of a speculative-empirical nature. They are not fully justified as yet by the relevant research results.

This shows that is is urgently necessary to develop such research, which would create a firm basis for the principles of an stimulating activity in the domain of science policy.

Principles of science policy as regards the selection of objectives

The principle of unity with socio-political ideology. The first fundamental principle of the selection of objectives of research work and developmental work relates to the necessity of basing it on the ideological premiss of the general policy of a given country. This policy might be orientated either for peace or for war, either for democratization and humanization of life or for the achievement and perpetuation of supremacy by some social group regardless of the interests of the working people and the nation as a whole.

In accordance with the principles of socialist development, it is assumed for a country like Poland that the ultimate goal of research and developmental activity is a full satisfaction of the living, cultural and social needs of man under conditions of peaceful coexistence with other countries.

Hence, the fundamental principle of the selection of goals for research work and developmental work is their adaptation to the material and non-material needs of man, broadly understood. The achievement of those goals should guarantee a harmonious and full development of men and their personalities. This principle might be briefly called the principle of unity with the socio-political ideology.

The principle of unity with economic policy. The second principle of the selection of objectives of research and development pertains to the necessity of their unity and kinship with the economic policy of the country.

From the general economic point of view, research and developmental work is tantamount to the preparation of future production and services, as also effects in the non-material sphere. The economic policy of a country must lay down proportions between expenditure on preparation for the nearer and the more distant future. This is connected with the problem of setting limits to research and developmental activity, expressed in the volume of expenditure earmarked for this purpose in relation to the overall state budget.

When international comparisons are made, expenditure for research and developmental work is usually expressed as a percentage of the national income. Quite apart from the not always comparable methods of the estimation of the national income itself, it must be pointed out that on the international scale this index constitutes no more then a measure of the relative burden to be borne by the people in the form of expenditure for science and technology. It appears from data for the last few decades that this proportional burden is increasing steadily, at a more rapid rate than the increase of the national income itself. This is illusirated by data supplied by UNESCO. [2]

This considerable rate of increase of expenditure on research and developmental work in countries on different levels of economic development, of different sizes and adopting different systems indicates how great a weight is attached everywhere to the role of science and technology in the shaping of the future.

The most authoritative index of the development of research and developmental activity is not the proportional share of expenditure in relation to the national income but the absolute sums earmarked for this purpose and their annual proportional increase. In this respect, too, the rate of increase of this expenditure shows a marked upward tendency. Setting aside the fluctuation of increments in certain countries caused by a war policy or space research, it may be said that in the course of the last two decades the expenditure on research and development in medium-developed countries was doubled every 7 to 8 years. This means an average rate of increase about 2.5 times higher than the national income increments in those countries.

The actual possibility of increasing the scope of research and developmental activity and the corresponding selection of material and non-material objecives is less than the average annual increments of expenditure. This has two causes. First, we note on a world scale a systematic decrease in the purchasing power of money, fiuctuating in different countries. Second, the costs of research and developmental work are increasing because of their ever greater complexity, as well as the necessity of overcoming increasingly great exploratory and technical difficulties. This is accompanied by an increase in the proportion of risk, which means that the costs of failures in exploratory activity are rising.

Taking all this into account, it may be estimated that in the course of the last two decades the average annual growth of the objectives set for research and developmental activity did not exceed from 8 percent to 10 percent every year. Can we consider this rate of growth of objectives as correct? Thorough economic studies and research on an international scale are indispensable here in order to ascertain scientifically justified grounds for determining the dimensions of research and developmental activity.

The relationship between the selection of goals of research and developmental activity and the economic policy of a country is also a result of the need to harmonize them with the programme of planned changes in the structure of production and services, as well as with the investment programme.

And last but not least, a few words should be said about the role of the analysis of the effectiveness and profitableness of tasks connected with the objectives proposed.

Recognizing the need for such analyses as an essential element affecting decisions on selection, it should be emphasized that analyses of this kind must:

— take into account social effectiveness — that is, for example, in the case of production, the advantages not only for the producer but also for the user;

— take into account not the present but the future relation of prices, which may be affected by, among other things, the undertaking that is being considered.

The principle of harmony of research and developmental activity in the material and non-material spheres. The next principle of the selection of goals of research and developmental activity relates to the establishment of correct proportions between material and non-material objectives.

There are so far no research results and theoretically motivated premises to justify the quantification of these proportions. This is why in most cases they are established on the basis of a comparison with other countries, and are shaped in a given country in relation to the situation there existing. The proportions are usually estimated by expressing them in the form of expenditure on work aimed at material and non-material objectives. The statistics of highly- and medium-developed countries do not differ excessively in this respect and indicate that between 8 percent and 15 percent of the overall expenditure on research work and developmental work are earmarked for the achievement of non-material objectives (cognitive research, research connected with the development of culture and education, etc.). For the time being, research aimed at non-material objectives absorbs as a rule less expenditure. [3] For this reason, the above-mentioned volume of expenditure means that from 25 percent to 33 percent of the overall time potential of scientific staff is connected with work on non-material objectives.

The situation in this field in Poland is similar, since some 12 to 14 percent of the overall expenditure on research and developmental activity in the years 1971–1975 were to be spent on cognitive research and the development of sciences connected with culture and social ideology.

A characteristic feature of the volume of work aimed at non-material objectives in Poland is the close relationship between the numerical strength of senior scientific staff in higher education (63 percent of the overall strength) and academic institutes who represent jointly about 70 percent of the total number of senior scientists. Cognitive research is conducted mainly in higher schools and Polish Academy of Sciences institutes.

The proportions of the material and non-material objectives of research and developmental activity should be based on premises connected with the fact that as regards scientific cognition science is but one and world-wide. True, various national schools emerge and develop in certain scientific disciplines in the individual countries; their sum, however, goes into making but a single world science. Contributions may be made to it by individuals of outstanding intellect, who are rather evenly distributed among the nations. It is apparent from this that wherever technical equipment does not play a decisive role, even small countries have a chance of world-scale achievements.

On the other hand, however, the results of world science sooner or later become accessible to all. The smaller the country, therefore, the weaker its potential and the lower the level of its economic development, the smaller must be the proportional share of its cognitive research in the overall research and developmental activity.

Scientific quests after a correct method of establishing the share of research the objectives of which are non-material should, though difficult, be considered to be an urgent responsibility in the domain of theoretical foundations of science policy.

Specificity of the selection of objectives for activity in material sphere. Of particular practical importance in the selection of objectives of research and developmental work is a careful establishment of principles specific for material and non-material objectives.

According to the thesis of harmonious development, the selection of material objectives should be based on the principle of giving proper consideration to objectives in all fields of human needs and endeavour.

The selection of objectives in the material sphere should be made with due consideration to problems and things. This means that in certain fields of human needs and endeavour, sufficiently comprehensive and complex to be considered as 'problems', we should indicate certain products, production methods, methods of the organization of work, and new kinds of services to be either invented, devised or perfected.

It is essential to name competent decision-makers as regards the selection and approval of objectives. We do not as yet know any method of selection based on premises of a mechanized, much less automated, system. This is why selection is made on the basis of conventional, compromise and synthetic proposals of teams of specialists of various kinds.

As regards objectives in the material sphere, teams of this type should be composed of scientists in appropriate disciplines, technical specialists, workers in the economy and managers. Proposals prepared by such teams should be received and approved (with appropriate changes resulting from confrontation with the economic policy of the country) by special organs of state authority.

The organization and practice of the selective activity in the individual countries do not greatly differ. Decisions are made by state executive organs of scientific policy with the rank of government commissions, committees or ministries. In Poland, this function has been entrusted to the Committee on Science and Technology which makes its decisions in consultation with the Planning Commission.

Specificity of the selection of objectives of activity in the non-material sphere. The selection of objectives in the non-material sphere of research activity has some analogies but also some specific characteristics as compared with the material sphere.

The most essential difference between the non-material and material spheres consists in the fact that selection in cognitive activity — that is, in activity the aim of which is to increase the store of knowledge, and thus to increase the influence exerted on the entire range of the non-material sphere — boils down in principle to the establishment of the directions of research activity and its tasks. Only to a certain extent is it possible to conduct work on cognitive problems (this being most frequently interdisciplinary research), in which the cognitive objective is set in advance. But even then cognitive objectives cannot be defined as concretely as objectives for research and developmental work in the material sphere. For in most cases they are objectives that are still to be discovered, which means that the degree of uncertainty and risk is here much greater.

Visible efforts have been made in recent years in research activity in the non-material sphere to increase the proportion of problems in which the cognitive objectives are obvious. In Poland, for example, the Polish Academy of Sciences has set several score of problems, a large part of them interdisciplinary. The biggest part of cognitive research, however, continues to be conducted in the field of the development of individual scientific disciplines.

Nearly all European countries have accepted the principle that cognitive research should be done in all scientific disciplines. [4] The principle is justified by the fact that even if a country's achievements will not enrich world knowledge, the conducting of research of a cognitive type (called also basic research) is necessary for educating a staff of scientists and specialists capable of understanding the latest and best achievements of science in other countries.

The principle of the fullness of cognitive research should be enforced together with the principle of the priority of the importance of a given research scheme (called also the principle of hierarchy) arising out of the needs and interests of a given country.

The majority of developed countries give priority to cognitive tasks belonging to newly appearing sections of the individual scientific disciplines, particularly of those which are building foundations for the future development of technique and production. These are domains in which science really enters the material sphere as a direct productive force or a force shaping production conditions. Generally included among such domains are, for example, selected sections of physics and chemistry connected with the development of microelectronics, certain sections of logic and mathematics connected with the science of computers, sections of sociology and psychology connected with the formation of a classless society properly utilizing all opportunities created by the development of science and technology, selected sections of biochemistry and biology connected with the development of living organisms together with a certain number of other disciplines in the domain of basic sciences.

It is generally recognized that the decision-making functions of state organs in the domain of cognitive research are limited to decisions on the directions of basic research and the selection of certain cognitive problems, especially those that are connected with research and developmental work in the material sphere. On the other hand, the selection and determination of tasks in the field of the development of scientific disciplines are the privilege of teams and individual scientific specialists.

Principle of the selection and hierarchy of objectives. In view of the limited means (specialists, funds, equipment, investment possibilities), a principle of successive selections is observed in both groups — that is, in the material and non-material spheres. In most countries, the selected problems are sub-divided as to hierarchy according to their importance and the range of the objectives achieved, the effects of research in hand, and the scope of their share in the implementation of planned tasks. The purpose of this hierarchy is the establishment of the degree of priority in the allocation of means, facilities and assistance in the conduct of work. It is also recognized as a rule that the highest rank in the hierarchy is given to schemes included in national plans. The place of individual schemes in the hierarchy is also decisive as to the scope of assistance rendered and supervision exercized by appropriate — especially national — organizations.

Poland has introduced a three-grade hierarchy of research and developmental work covered by the state (central) plan, ministerial plans and branch plans.

The principle of up-dating objectives, directions and tasks. The last principle of the selection of objectives for research and developmental work is a periodical revision of the goals, directions and tasks. This is done as a rule once a year. This could be called the principle of a 'striding' selection, since it involves not only amendments in the selection already made, but also the setting of new objectives and tasks, so that we may always have before us a proper time horizon indispensable for the implementation of time-consuming research and developmental work. It is assumed in Poland that this time horizon should be not less than 5 years. This means that objectives for at least five years ahead should be brought up to date every year.

Ensuring the implementation of tasks and achievement of objectives

Scientific and technical cooperation with foreign countries. Research and developmental work the results of which are the achievement not only of national scientific and technical institutions but also of such abroad are the source from which are met the country's needs as regards progress in economy and the non-material sphere. Every country has as a rule more needs than it can satisfy with its own forces. The first principle of the implementation of tasks and the achievement of objectives of research and developmental work is, therefore, cooperation with other countries as intensive as possible and widely conceived, and the best possible utilization of foreign scientific and technical achievements.

This principle may be particularly well justified in the case of medium and small countries. As regards Poland's demographic potential, for example, her population constitutes about 1 percent of the population of the Earth, her economic potential is estimated at about 2 percent of the world potential, while her place as regards her intellectual possibilities, taking into account among other things scientific and technical literature, is judged to be between 8th and 10th in the world list. These figures indicate that if we are to keep in step with the most highly developed countries with the biggest potential we cannot count on our own forces and possibilities alone.

We are beginning, therefore, to pay increasing attention to the problem of cooperation with foreign countries. We have at present the following classification of the types of such cooperation:

(1) execution of joint scientific schemes and research and developmental undertakings;

(2) import and export of non-materialized 'scientific and technical know-how' — that is, exchange of information and documentation, advisers and experts;

(3) import and export of 'materialized technical know-how' — that is, licences together with production equipment and assistance in the practical application of the latest achievements to production or services;

(4) training and perfecting of specialized personel: Poles abroad and foreigners in Poland;

(5) use of international cooperation conducted within the framework of international governmental and non-governmental organizations.

Considerable differences may be noted in the to-date scope, organizational standard and effectiveness of scientific and technical cooperation in the individual countries. It may be said on the basis of the situation in Poland and in other comparable countries that cooperation with countries abroad is not conducted so far in a complex and all-round manner.

Hence the science policy in the individual countries must be more energetic in this domain. It is also necessary in the interests of humanity to counteract excessive commercialization of cooperation with countries abroad. Normal trade exchange must not be curbed by various bans of the of 'embargo' type which in disregard of the interests of numerous nations are extended to many most of the up-to-date designs and achievements. It seems desirable to conclude a world-wide convention on international scientific and technical cooperation.

The programme adopted by the 15th Plenary Session of the Council for Mutual Economic Assistance in Bucharest in 1971 may serve as an example of a perfectly organized multilateral scientific and technical cooperation. The programme concerns directly the C.M.E.A. member countries, but it is open to accession by any other country.

Strong emphasis must at the same time be laid on the great possibilities of bilateral cooperation which still remain unnexploited.

A particularly important element of scientific policy in the sector of cooperation with foreign countries is the creation of proper atmosphere around the adaptation of world achievements.

Every country, including Poland, has the social right and duty to assimilate the achievements of others. This is justified by the fact that with her own creative work she contributes to the achievements and development of scientific knowledge which is recognized to be the good of the entire mankind.

This is connected with the problem of how to treat achievements of an original character and achievements of what are called a reproductive character. There is every justification for social respect and various forms and expressions of recognition — including financial recognition — of original, national work. This, however, must not serve as an excuse for discriminating reproductive adaptation work. Quite the contrary, it is in the social interest to evaluate achievements not on the basis of their originality but of their social usefulness, up-to-dateness, quality and cost. The practical experience of several countries, one of which is Japan, where rapid adaptation has in recent years been regarded as an important element of science policy, indicates that it is desirable to treat this kind of trend as a stable and universal principle.

This holds also for Poland, where the degree of the exploitation of foreign achievements by way of adaptation is so far insufficient. The situation is similar as regards the need to expand commercial forms of cooperation with foreign countries, particularly by the purchase of licences. The funds spent in Poland for this purpose constitute a much smaller proportion than in other European countries of the overall expenditure on research and developmental work.

One other essential question appears, important from the point of view of the effectiveness of such cooperation. Cooperation with foreign countries should be treated as an integral part of the execution of national tasks. This means that such cooperation should include tasks appearing in the programmes and plans of the cooperating countries. The observance of this principle on the part of every country is an expression of the practical introduction of the principle of cooperation on a scale which is interesting and necessary for all the cooperating countries.

Planning research and developmental work. The second principle of science policy in the implementation of tasks and achievement of objectives of research and developmental work consists in the recognition of planning as the main stimulator of scientific and technical activity.

Planning constitutes an integral part of the management and organization of research and developmental activity. It entails operations which might be divided as follows according to the degree of their accuracy and obligatoriness:

— preparation of prognoses — that is, the forecasting of mutual relations between developmental processes and their effects, without assuming obligatoriness of actions and strict deadlines;

— preparation of programmes which also contain elements of prognoses, but contain already actions necessary for the production of planned changes and the achievement of objectives, together with a rough estimate of the time and funds needed;

— preparation of plans which contain a definition of objectives necessary for the implementation of tasks, dates of implementation, indispensable means and contractors.

This means that this plan closes the final stage of the planning processes, and has the highest degree of obligatoriness of all documents of planning. The time period covered by a plan is as a rule shorter than the period covered by a programme, while that covered by a programme is shorter than that covered by a prognosis.

If the planning process and the prognoses, programmes and plans prepared in the course of it are to serve their purpose and be really useful, they should meet certain requirements.

The first of these is proper differentiation and adaptation of planning to the specificity of a given kind of work. Without going too deeply into the technique of the preparation of planning documents, it is necessary to underline the specific features of planning in the field of research and developmental work as compared to economic planning which developed earlier.

In view of the unavoidable risk and degree of uncertainty always attendant on research and developmental work, the accuracy of planning has to be less — that is, the tolerance of time limits and even the tolerance of funds and requirements pertaining to the effects obtained should be greater than in economic planning. The interpretation of this principle should depend, however, on the nature of the research and developmental activity.

Accepting this point of view, we distinguish three kinds of activity:

— work with a complete developmental cycle — that is, stretching from research to practical application and including all tasks indispensable for the achievement of the ultimate objectives which is as a rule utilitarian, socially useful;

— work with an incomplete developmental cycle, by which we understand cognitive work, reconnaissance work (trial work), expertizes, analyses, etc., from which no complete developmental cycle is required;

— general technical activities, by which we understand work supplementing or assisting research and developmental activity, especially with a complete developmental cycle. These include scientific and technical information, informatics, standardization (in a broader sense of the term), additional training, popularization, etc.

As regards work with a complete developmental cycle, it is necessary to have complex coordinating plans worked out with accuracy sufficient to synchronize them with economic plans when it comes to the practical application of their results in the economy. It is usually assumed with regard to long-range plans that they should be accurate to within one year as concerns their finalization, and to within one quarter of a year as concerns the start of a cycle. This does not apply to the operative plans for the individual tasks where accuracy may be correspondingly greater.

As regards work with an incomplete developmental cycle, plans may be very highly simplified. They may not be prepared at all if the tasks are short-term and not very complex. A similarly simplified approach may be adopted as regards plans of general technical activities, especially if they are not repeatable.

Every scientific and technical agency may, and even should, pursue these three kinds of activity. On the other hand, depending on the character of the agency and its temporary engagement in certain research and developmental activities, the proportions of the amount of time and funds expended on these activities may change.

In Poland, there are some fixed proportions for the individual kinds of activity for agencies of different types. It has been accepted, for example, for scientific research institutes operating in the sphere of national economy that work with complete developmental cycles should absorb not less than 50 percent of their working time, general technical activities not more than 30 percent, while the remainder should be used for work with an incomplete cycle, with the reservation that cognitive work should absorb not less than 5 percent.

Analogous fixed proportions have been established for higher education units and for the Polish Academy of Sciences.

The immediate superior authority is in all cases empowered to establish every year individual proportions, even differing from the fixed ones, if justified by social and economic advantages.

Another requirement of proper planning is its being brought

up-to-date with a constant — usually 5-year — time horizon. This is usually done at intervals of one year. One of its purposes is the erasure from the plans of objectives and tasks that cease to be justified, either as a result of development in other countries or the unfeasibility of obtaining a positive effect. Another purpose is the introduction of new objectives and tasks arising from new needs or possibilities of science and technology. Further purpose is to increase the strictness of the obligatoriness of tasks as the date of their finalization approaches. This pertains especially to work with a complete developmental cycle.

The bringing up-to-date is connected with the third requirement — namely, proper division of functions and powers in planning. Objectives, deadlines and funds should be fixed by appropriate decision-makers in consultation with teams of specialists performing the work, while the selection of tasks and methods indispensable for the achievement of the objectives agreed upon should be left to the specialists concerned. In case of necessity, they should have the right to change the 'road to the objectives' without consultation with the decision-makers — provided that the actual objectives are not changed — put the deadlines forward and increase the expenditure.

Another requirement of rational planning is the division of work into specialized work demanding the participation of scientific and technical specialists and strictly planning work which may be performed by a suitable trained administrative personnel.

Of particularly great importance among the various types of plan are complex coordinating plans of complete developmental cycles, corresponding in Poland to what are called 'key problems'. For the coordinating plans contain the structure of tasks deemed indispensable for the achievement of the objectives set, their distribution among all the cooperating institutions, together with a suitable calendar and list of means for implementation, as well as the name of the institution responsible for the whole of the problem.

In this way, coordinating plans are an external expression of the complex approach to the problem, and at the same time an instrument in the participation of scientific and technical institutions in the selected cycles of the research and developmental activity.

Finally, an extremely important requirement of plans is that they must guarantee the implementation of tasks and the achievement of objectives set. This means that the plans must guarantee all expenditure and means (financial, personnel, equipment and organization) indispensable for the achievement of the objectives set. This is in Poland more and more universally called the principle of 'unity of tasks and means'.

Financing of research and development work. Another principle in the field guaranteeing for the implementation of tasks and the achievement of objectives in research and developmental work is the recognition of the fact that financing is — besides planning — the most effective method of stimulating scientific and technical activity.

This may be achieved through proper financing earmarked, according to a previously considered motivation, either for certain objectives and tasks, which in the language of economists is called objective financing, or for the activity of whole institutions or branches without differentiation into objectives or tasks — subjective financing.

Throughout the world may be observed a tendency to increase the share of objective financing not only as regards complete developmental cycles but also as regards cognitive work, where it is called financing by subsidy for certain tasks.

Nowhere, however, is any tendency observed to discontinue subjective financing. It is reserved mainly for institutions or groups conducting cognitive research in which it is difficult for any length of time not only strictly to define the goals of research, but even to select concrete subjects. Free-of-charge general technical activities may be treated in a similar manner. A variety of subjective financing is also the subsidizing of the individual activity of scientists whose objectives and the confidence they enjoy as a result of their work justify financing without unnecessary argument and strict control of the spending of the subsidies. This form of financing is used, however, in exceptional cases only.

Objective financing in Poland in the years 1971–1975 was planned to cover about 65 to 70 percent of all expenditure on research and developmental activity.

Proper exploitation of the positive aspects of objective financing requires a suitable form of financial economy. It should enable every institution, irrespective of its character and the nature of its work, to benefit from both objective and subjective financing. It is necessary to emphasize this principle, since economy based on principles of self-financing — that is, on economic settlement of accounts — is sometimes wrongly identified with objective financing, while subjective financing is identified with economy in the form of a budgetary unit. Self-financing should not exclude the possibility of obtaining subjective subsidies, while subjective financing may be used also in case of economy in the form of a budgetary unit. Self-financing should not exclude the possibility of obtaining subjective subsidies, while economy in the form of a budgetary unit should not exclude the possibility of obtaining payment for work on a certain subject.

The above principles, postulated by the scientific and technical base, require the introduction of suitable reforms in many countries, including Poland. It is necessary to try to work out a typical financial economy, uniform for all institutions of the scientific and technical base, and permitting at the same time, depending on the individual situation and the character of the institution, the fixing of correct proportions between objective and subjective financing.

Simultaneously, it should be recognized as justified to increase the proportional share of institutions operating on the principle of self-financing mainly through the conclusion of contracts with units of national economy and through the acceptance of orders from authorities or institutions supporting research and developmental activity.

For such form of financial economy establishes very effective ties between scientific and technical work on the one hand and the needs of the economy on the other; it increases the sense of self-sufficiency of an institution, and it creates at the same time its own financial base for various kinds of material incentives for its staff and for the institution as a whole.

Experiments made in Poland in this domain in 1969 and 1970 show positive effects of this form of management in units operating in the sphere of economy. It is planned, therefore, to popularize this form in future.

Countries with a planned economy introduce certain limits for some elements of financing — e.g., for wages fund (based on the amount of money on the market in relation to market supplies), investment expenditure (depending on the building potential), or currency expenditure (aimed at preserving the balance of payments). These measures — justified on the scale of a state — are not always indispensable on the small scale of individual organizational units or even whole groups of scientific and technical institutions. For in such cases the overall expenditure may be treated as fully exchangeable for the equivalent expenditure of all kinds, since this constitutes no threat to the economic equilibrium of the country. Quite the contrary, it increases the sense of responsibility of the administration of the given institution and makes its activity easier by cutting down indispensable formalities.

In Poland and some other socialist countries, studies are in progress and preparatory steps taken for changes in this direction.

Securing the development of the potential of the scientific and technical base. Other elements guaranteeing the implementation of tasks and the achievement of objectives of research and developmental activity — namely personnel, investments, equipment, organizational structure and increase in the number of institutions — should be considered not only from the point of view of their adjustment to the already planned, envisaged work, but also from the point of view of creating a potential for a more distant future, which would permit a further, correct development of research and developmental activity.

Of these two objectives — namely to ensure the implementation of current and planned tasks and to form at the same time a suitable potential for the more distant future — more space and attention are devoted in science policy to problems of creating the potential.

Ensuring the implementation of tasks already in hand or planned boils down mainly to the exploitation of potential already in our possession. It remains, therefore, solely in the sphere of activity of the managements of the appriopriate institutions. The intervention of superior authorities is mostly of the character of sporadic assistance, etc.

If, therefore, science policy is dominated by the need to create a future potential of the scientific research and technical development base, then this policy must be based to a large extent on prognoses of orientation and proportions of the development of research and development institutions in various branches of material and non-material activity. And it is precisely that this constitutes the main principle of creating the potential of the scientific and technical base.

We have as yet no generally accepted prognostic methods nor, therefore, any generally accepted methods of programming and planning the development of the potential of the research and developmental base. Consequently, its development is not always sufficiently harmonized with the subsequent needs in the domain of research and developmental work. It happens sometimes that in certain domains the potential of the base is a factor limiting possibilities, while there are also cases in which the potential shows reserves not full exploited for the country, which gives us an opportunity to develop the export of 'technical know-how'.

These phenomena may be observed also to a certain extent in Poland, a fact which has given an impulse to the undertaking of deeper studies and work on a proper orientation and development of the potential of the scientific and technical base.

Irrespective of the future results of this work, it is possible already to outline some quite popular views concerning the various component parts of the potential — that is, personnel, equipment and premises, as well as the number and organizational structure of the institutions of the scientific and technical base.

Development and use of the personnel potential. Principles in handling the personnel problem are particularly important. Personnel is an element of the general potential of the scientific and technical base, and it must be handled, therefore, in accordance with the principles of the science policy. But the personnel includes also the makers and executors of that policy. Problems of the scientific and technical base, though they possess a high degree of their own specificity, constitute also a part of the greater problem of personnel in general, for which we use also the term of 'personnel policy'. It is possible, therefore, to present the general principles of the personnel policy, which are also generally binding. But this would be beyond the scope of the present paper. Only some more specific problems and principles of the creation of the personnel potential of the scientific and technical base are touched upon below. This does not pretend to be an exhaustive presentation of the issue.

When it comes to increasing the number of persons employed in the scientific and technical base it is necessary properly to differentiate their functions and qualifications, and to distribute them among the various kinds of activities and types of institutions.

As regards institutions already in operation, in most cases we use for shorter periods the extrapolating method, most frequently of a proportional increase in employment, which is as a rule the result of the absence of a well-grounded view on the shaping of employment increase. Such a method cannot be considered correct in the long run, and must be replaced by prognosing the development of the various branches of economy, culture and science, and the resulting demand for the development of the personnel potential of the appropriate institutions. For only in this way is it possible to introduce desirable changes in the structure and distribution of personnel.

The general well-known principle of observing a suitable proportion between auxiliary personnel and creative personnel must be adjusted to the individual type of the institution and the type of the activity. Data supplied by institutions whose personnel is satisfied with the existing proportions of employment and whose working results are good show that the above-mentioned proportions fluctuate between 3 : 1 and 42 : 1.

Some institutions, however, show marked deviations from correct proportions. An indication of this are first and foremost difficulties in the performance in their work, an insufficiency of which is expended on some of the stages.

In Poland, for example, erroneous proportions in the employment of auxiliary and creative personnel may be noted in some higher schools. This justifies the remedial steps already taken.

The distribution of personnel in the institutions of the various sections of the scientific and technical base should depend on the organizational conception of the base as a whole. In the United States and the Soviet Union, for example, the majority of scientists are employed by institutes and institutions working in the economic sphere. In most European countries, the majority of scientists work in higher schools and institutions connected with them.

This group of countries includes Poland, where about 63 percent of senior scientists are employed in the educational system. An analysis of this situation shows that the research institutions of the national economy employ too few scientists. This is a result of too small a rate of increases in the number of scientists directed to those institutions. The improvement of this situation in Poland is already an object of systematic organizational efforts of the appropriate authorities, and will continue to receive their attention for a long time to come. This is due to the following factors. We were on the threshold of producing about 2500 young doctors of sciences every year up to 1975. In view of the still very considerable needs of higher education, only some 40 percent of them can be employed in the economic sphere. Such a distribution will gradually change the proportions so far in the distribution of scientists.

Another problem of personnel policy are the qualifications of that personnel. Obviously, the qualifications of all employees of the scientific and technical base must be high. But special requirements must be met by the creative personnel. They should include the group of scientific workers and a part of the scientific and technical workers. An analysis of the work of these personnel groups made every few years shows that only a certain proportion of employees achieves actual creative effects. According to unofficial opinions exchanged at the MINESPOL conference of ministers responsible for the scientific policy in European countries the proportion fluctuates in certain of those countries from 50 to 75 percent. This means that science policy must operate more and more energetically towards the development of creative activeness.

Steps in this domain, either undertaken or being prepared in Poland, reduce to:

— working out a system of 'catching young talents' still in the course of education and ensuring their proper care, guaranteeing rapid development and employment in accordance with their abilities;

— developing a system of supplementary training for research and technical workers, both as regards the methods of work and the supplementing of their store of knowledge with the latest achievements of science and technology;

— working out and introducing a new system for the evaluation of scientific and technical workers, based on essential criteria: namely, the effects of their creative work;

— introducing a system of selection and rotation of employees in the creative group; this offers practical difficulties because of the simultaneous striving to produce a sense of stability of employment, highly desirable in creative work;

— improving the system of remuneration, incentives and material and moral rewards for creative workers;

— shaping and deepening an attitude of dedication to, and love of, creative work, quite natural in scientific and technical workers;

— introducing various types of official and organizational privileges (e.g., release from certain formal responsibilities of discipline of work) for scientific and technical workers distinguishing themselves by their systematic and fruitful creative activity;

— relieving creative workers from work of an administrative type, by employing a sufficient of properly qualified administrative personnel.

Each of the above directions of activity has its own problems to be appropriately dealt with. They deserve separate treatment and a broader presentation than is possible in the present paper.

Increase of research and technical equipment and its exploitation. A sufficiency of up-to-date research and technical equipment multiplies the possibilities of research and permits a proper use of the creative talents of the staff; at times it is even quite indispensable for the performance of certain tasks. The correctness of this thesis is confirmed again and again. It is necessary, therefore, for science policy to pay special attention to problems connected with the supply of research apparatus and technical equipment for the scientific and technical base.

A generally recognized principle in this domain on the scale of the whole of the scientific and technical base is a gradual increase of the technical implements of the staff in the basic group. This means that the proportional increase in the net value (after the deduction of depreciation) of the research and technical equipment should be greater than the proportional increase in employment in the basic group.

On the scale of single, especially small, institutions, it is admissible to depart from this principle for short periods. It should be respected, however, over long periods.

With the exception of a few countries with a highly developed industry producing research apparatus and equipment, all the other countries base their equipment supplies on import. Poland, for example, imports about 60 percent of her research apparatus. Hence the problem of increasing research equipment is to a great extent a problem of increasing expenditure in foreign currencies. This is appreciated by the government. For the years 1971–1975, it has earmarked for this purpose sums enabling a greater than 50 percent increase in research equipment.

From the point of view of science policy, it appears necessary to try to develop the home industry of research apparatus and equipment, and to conclude appropriate international co-production agreements. As an example of such a venture we may cite the creation of an international institution called *Interatominstrument*, with headquarters in Warsaw. The development of home production, on the other hand, is stimulated by, among other things, the introduction of convenient conditions of work, and home and export sales for producers. The increasing value of apparatus and research and technical equipment systematically raises the rank of the problem of proper exploitation of these valuable instruments. Principles of the exploitation of production machinery and installations cannot be applied in their entirety to research equipment. Nevertheless, the universally encountered fact that unique apparatuses are utilized to a very small degree inclines us to intervene in this domain as well.

In this connection, Poland is taking certain steps, the most important of which are:

— organization of a regular information service on the possession and location of particularly valuable and unique equipment;

— support of the development of research services on both the national and international scale;

— setting up of and all-round assistance for joint socialist environmental laboratories for meeting the equipment needs of a certain number of institutions;

— control of the exploitation of particularly valuable apparatus;

— introduction of licences for the purchase of unique apparatus, subject to a guarantee of its being properly used;

— introduction of facilities for the sale of apparatus redundant for the actual possessor.

All these measures should lead to much better exploitation, especially of very valuable research apparatus. From the point of view of science policy, apparatus must be treated as a factor in specific scientific production second in importance only to the human factor.

Shaping the structure of the scientific and technical base. The creation of new institutions of the scientific and technical base and the transformation of the existing ones, with a view to establishing an organizational structure better adapted to the present needs, comprise one of the more difficult sectors of the scientific policy.

The problem is especially complex in Poland, where the construction of a modern potential of the scientific and technical base did not begin in fact until after World War II under conditions of tremendous shortage of specialized personnel. We were then also just beginning the construction of our economic potential. In those circumstances, we arrived at the concept — fully justified at that time — of setting up mainly what were called single-purpose institutions — that is, institutions specializing in accordance with the separate links of the developmental cycle, e.g., design offices, technological offices, investment design offices, research institutes, etc. These single-purpose institutions could work for the needs of a larger number of plants or even entire branches of industry.

But as the national economy developed, the requirements and needs of the various branches and plants began to increase, and the single-purpose institutions had to narrow down their activity.

This created the need to set up further single-purpose institutions working on comparatively narrow subjects.

At the same time, the demand for combined research and developmental operations made it necessary to begin an increasingly complex cooperation both between the institutions of the scientific and technical base and between the national economy and the scientific and technical base. This produced numerous difficulties and reduced the effectiveness of the research and developmental work. This, in turn, was the reason behind the crystallization of the present policy of stimulating the development of the scientific and technical base and its organizational structure.

The main theses of the present conception of shaping the structure of the scientific and technical base are as follows:

(1) wherever it is justified by the permanency and volume of cooperation with regard to a given objective (e.g., in the domain of the development of micro-electronics), it is desirable to try to integrate the individual single-purpose institutions, creating multi-purpose institutions capable of solving complex problems of a complete developmental cycle;

(2) single-purpose institutions, engaged for example in research, which have no suitable cooperators in the domain of the other links of a developmental cycle (e.g., construction and technological designing) should be transformed by way of evolution into institutions of the research and development centre type;

(3) when creating new institutions, preference should be given to the conception of a multi-purpose organization, of the research and development centre type.

These theses involve the principle of setting up institutions in accordance with their objective complexity. This is bound up with a principle according to which the scientific and technical base should have the closest relations possible with the corresponding branch of the national economy.

This means that institutions of the scientific and technical base may be linked with industrial plants or industrial federations or ministries. However, the organizational subordination arising from this must not liquidate or reduce the scientific independence and autonomy in the evaluations and proportions of development, which are indispensable for research and developmental institutions. For in the former case the role of a research and developmental institution may be narrowed down to a servicing function, which stands in contradiction to the fundamental striving for a maximum development of creative initiative, according to which the scientific and technical base should submit its 'offers' concerning development.

The above principle of organizational integration must not be regarded as fixed and invariable. For certain situations in the national economy fully justify the existence of single-purpose institutions. It should be understood, therefore, that the striving for organizational integration on the principle of objective complexity is intended to bring about a correspondingly greater increase in the proportional share of multi-purpose institutions, of the research and development centre type. However, single-purpose institutions will be permitted to operate at the same time in accordance with the possibilities and needs.

Another example of the further development of single-purpose institutions are experimental establishments and laboratories whose number should be even higher in order to strike a balance between on the one hand the scope of research and the appearance of new projects and on the other their practical checking by way of prototypes and semi-industrial technological installations.

The integration measures in the realm of the scientific potential of the higher school system are also justified in Poland, due to their having been until quite recently scattered as small and all but independent organizational units (chairs). A specific feature of this integration must be the observance of the principle of the unity of science, education and training. This must pertain not only to strictly higher school institutes but also to institutes organized jointly by higher schools and industry, which may be set up on the basis of long-term agreements between the national economy and the system of higher education.

The policy behind the development of the scientific base of the Polish Academy of Sciences has also some specific features. The setting up of Academy institutions is justified only where there is no corresponding unit of the system of higher education, or where it is necessary to concentrate research in a given domain in order to intensify it and to obtain quickly the results needed. On the other hand, institutions of the scientific base of the economy should be set up among the existing scientific institutions which as a rule carry out research in new developing fields. The forms of their emergence may differ: from using a part of their staff as a nucleus of a new industrial research institution, to transforming an entire higher school institution into an agency operating in the industrial sphere.

There exists, however, the danger that the principle of objective complexity in the integration processes might cause a further differentiation of the entire scientific and technical base through its being divided into various 'lines' linked with the separate sections of the economy, education, etc. In order to counteract this, therefore, we observe also at the same time the principle of the integration of the scientific 'lines'.

This is expressed in, among other things, the creation of facilities for transferring and loaning scientists between institutions in various 'lines', as well as for scientific and technical cooperation, in the working out of standardized economic methods, etc.

Proper regional distribution of institutions is an important problem in shaping the organizational potential of the scientific and technical base. This calls for consultation and harmonizing of the regional policy of the country with programmes of the development of the scientific research potential in the individual sections of the national economy. General principles of the regional development policy of the country are primarily applicable here. Worth bearing in mind among them is first of all the principle according to which institutions of the scientific and technical base are to be established above all in regions with the densest concentration of plants expected to apply and utilize the results of the research and developmental work in a given domain.

Utilization of the results of the research and developmental work

The importance of practical application and popularization. Numerous experiments made in a large number of countries show that it is not always the country where a scientific discovery or technical effect is achieved that reaps the first, if any, fruits of its achievements. It is frequently forestalled by countries possessing a stronger and more efficient apparatus for the exploitation of scientific and technical novelties.

There are also cases in which novelties mastered in practice are popularized slowly, which prevents us from feeling their positive effects of large-scale progress.

These two observations confirm the great importance of the practical application and popularization of the results of research and developmental work, as well as of novelties produced by inventions. Even the best selection of objections and tasks, and the best implementation of tasks and the achievement of objections, as well as the best possible conduct of research and developmental work, will not help if there are delays in the practical application and proper popularization of the results of that work.

This indicates the indispensability of the closest possible link between on the one hand the cycle of research and developmental work, on the other — practical application and popularization. Hence this problem must be also taken into account in science policy, even though practical application and popularization are mainly the concern of national economy units and not the scientific and technical base. There is no doubt, however, that a proper cooperation of the two partners is necessary in this field.

The experience of many countries, including Poland, shows that practical application and popularization are the least efficiently operating links of the developmental cycle. No efforts are being spared, therefore, to increase the efficiency of the methods of practical application.

Poland has also worked out a complex programme of activity which assumes that methods of improving the efficiency and effectiveness of practical application and popularization should concern both national economy units and scientific and technical base units, and that they should be of an organizational, planning, economic and moral nature.

Stimulating organizational-planning. Factors which may be classified as stimuli to organizational-planning include:

— the setting up of mixed research and application teams

composed both of the scientific workers and employees of the plants which are to use a given novelty in practice;

— strict confrontation of the plan of research and developmental work with the investment plan and the plan of innovations in production, which means that the applications foreseen should have a suitable guarantee in the investment plans, since the majority of the results of research and developmental work cannot be applied in practice without investment, greater or smaller;

— all investment undertakings should be based on the opinions given by suitable teams of specialists as regard their modernity, and only investments which receive a positive opinion in this respect — that is which receive a 'modernity clause' — may be permitted to be built;

— suitable orders for every item in the programme of innovations in industrial plants aimed at the introduction of new products, modernization or modification of a given product, should be placed with appropriate units of the scientific and technical base for research and developmental work;

— the inclusion of all new proposals resulting from successfully completed research and developmental schemes during the preparation of operative annual plans of practical application and popularization;

— the market distribution lists should contain new products of home production, which are a mark of the sufficient rate of the popularization of these achievements in this country, subject to the requirement that mass consumption products should be submitted to the opinion of the trade unions.

Economic stimulation. Economic stimulation as a supplement to the stimuli of organizational-planning may prove to be particularly effective.

Methods of economic stimulation embrace first of all those that effect directly not only the scientific workers who have new achievements to their credit but also the entire teams working on the practical application and popularization of these achievements. This means that bonuses should depend on a punctual and good performance of the tasks involved in application, and that prizes, provided for in advance, should depend on the effects obtained from a given achievement in a given period of time.

Another economic factor which should have a positive effect on the rate and quantity of practical applications is the renovation and modernization of production. Poland is trying to work out a system of what are called cascading prices — that is, prices which drop as the date of the introduction of a given product ages. It is proposed in this connection to introduce for 'novelty', 'normality' and 'obsolescene' separate prices involving decreasing profits. It will be unprofitable, therefore, for a producer to continue the production of old products; this should incline him to produce a new or modernized product.

And to prevent a simulated modernization and an unjustified rise in the production costs, more exacting technical requirements will be introduced simultaneously (A, B, C categories of technical standard and Q and I quality mark) for new products. It is also proposed to introduce the principle of comparability of prices with those of the strongest competitor in the international market. As regards mass consumer products, it is also desirable to establish a stricter relationship between the system of retail prices and the periods of novelty, normality and obsolescence.

It is planned also to introduce other stimulants of an economic nature aimed at producing an atmosphere of demand for technically advanced products. As a case in point we may cite the introduction of a different method of calculating labour costs in the estimate of the costs of new application (of the volume of the general consumption fund), which under Polish conditions means an increase of the share of labour costs by about 30 percent. This should have a positive effect on lowering the limit of the profitability of innovations, especially connected with automation which usually yields economies in labour.

Moral stimulation. As regards the moral stimulating of employees responsible for the rate and success of new applications, the Polish programme includes the following principles:

1. A new application should be treated as an integral part of the preparation of production; it must, therefore, have incentives and sanctions analogous with those in production. This means that the

managements of industrial plants and service establishments, as well as the respective personnel, should be held responsible for lack of planning and non-execution of new applications, and the sanctions should be the same as in the case of production tasks. This principle of responsibility for technical progress should be extended to all levels of industrial authorities.

2. Their share in the introduction of new applications is the sole criterion of the evaluation of scientific and technical staff.

3. Outstanding scientists and employees making a major contribution to the practical application and popularization of socially and economically important innovations should be given public recognition, be popularized and recommended for suitable decorations.

Undertakings aimed at improving the processes of new applications and popularizations are connected also with the necessity to improve the method of the determination of the effectiveness and profitableness of research work and new applications, consisting in the introduction of a principle, already referred to, of the social effects of a given novelty as concerning both the producer and the user. For it sometimes happens that certain innovations are of actual advantage not for the producer but precisely for the user.

Where works of a purely scientific character are concerned, the term 'new application' must be suitably modified. It may be replaced by the utilization of research results in creative technical work. Regarded as a 'new application' may be the results of theoretical studies and research on the anisotropy of the strength of plastics, if used in construction designing which has yielded economies in materials and an increase in the safety coefficient of a construction.

Generally speaking, problems of new applications are connected organically with the process of reproduction, that is with the general system of the operation of our economy. It may be assumed, therefore, that effective improvement of the system of new applications, and the resultant increase in the effectiveness of research and developmental work, should be connected with reforms in the system of the economy as a whole.

Added in proof. The article mainly deals with the principles of science policy and their motivations. In many places it is also shown how those principles influence the system of science and technology in connection with the operation of national economy, in particular of that of Poland. Some information as regards programmes of activities in the sphere of science and technology in Poland is included as well.

Though in the period between the preparation of the article (1970) and its present appearance in print the science of science, which is one of the main theoretical foundations of science policy, has greatly developed, the principles of science policy, presented in the article, have not been changed. The examples, however, referring to the system, i.e. the structure of the potential of science and technology and their relations with the national economy, and in particular the programmes of activities in the field of science and technology, as well as the utilization of obtained results in the national economy, have undergone many changes. Therefore, certain bits of information and figures should be regarded as alredy historical data.

NOTES

[1] Further on, the arguments will relate chiefly to scientific policy as regards the achievement of goals in the material sphere and as regards increasing the store of knowledge (scientific cognition). The principles in question may be adapted *per analogiam* to science policy as concerning the achievement of goals in the entire non-material sphere.

[2] Reports to the 1st MINESPOL Conference of Ministers for Science and Technology of European Countries.

[3] It must be pointed out that it is not only expenditure on cognitive research that must be taken into consideration. For this work requires to an ever greater extent an abundance of research apparatus and technical equipment. Cognitive work in some fields is becoming, therefore, increasingly expensive. As an example here may serve research on the structure of matter and energies of inter-molecular bonds.

[4] Result of the Conference of Ministers for Science and Technology of European Countries — June 1970.

Wojciech Gasparski

SYSTEMS APPROACH TO RESEARCH IN THE DOMAIN OF THE SCIENCE OF SCIENCE (METHODOLOGICAL REMARKS)*

Translated by Witold Dreger

Introduction

An attempt at defining the methodological assumptions of complex research on the development of science and technology, that is of research approached in a systemic way, should be at first preceded by the formulation of the initial object of research and of the resulting definition of science as the object of research thus delineated.

The second task consists in establishing the essential problem area of research, i.e., the first order decomposition of the problem of research on the science of science into relevant subproblems.

The third task consists in the decomposition of these subproblems into subproblems of a lower order until such detailed research problems are attained which are solvable by known detailed methods, called also 'research techniques'.

The fourth task, covered already by the scope of the organization and coordination of the course of research on science, consists in the distribution of particular research problems among individual research institutions which are to solve the problems in question with the use of the detailed methods ('research techniques') proper to them.

The fifth task consists in the aggregation of the results obtained in the course of solving research problems into the synthetic

---

* First published as 'Badania naukoznawcze w ujęciu systemowym in *Materiały i Informacje*, Zakład Prakseologii Polskiej Akademii Nauk No. 1 (1972).

result — the solution of the initial problem of complex research of science and technology.

Taking into account, that the history of the complex research on science is short and that the methodology of the systems research is likewise in the initial stage of its development, we see that the study of the above-mentioned essential problems must be coupled with the research on the methodology of the systems research. For that very reason it would be worthwhile to precede the discussion of the above-mentioned problems with a short review of the development of this methodological conception of scientific research.

Origin and development of the methodology of systems research

The modern specialists of the methodology of science unanimously stress that contemporary science, apart from ascertaining *a posteriori* the interrelations between the studied objects, searches for these relations — due to the complexity of the studied object — in the process of identification of an *a priori* formulated theoretical model. The most apt expression of this trend is the terse description of the cybernetic procedure derived from a saying attributed to Picasso:

> *D'abord je trouve, je cherche ensuite.*
> ("First I find it, then I look for it".)

As is usual in such cases, it would be rather difficult to pinpoint a particular moment in the history of science which should be counted as the starting point of the acceptance of the *a priori* methodological trend in research concerned with the reality (that is other than, for instance, mathematical research). It seems, however, that — otncontesting earlier attempts in this direction — the discussed methodological trend began to spread only when the need for the scientific changing of the reality was appreciated. That is, when besides the positive research concerned the prospective research, known also as the applied or practical one, were undertaken aimed at changing reality.

The changing of reality, which at the information level (i.e., this level on which information is the processed subject-matter), takes the form of design or planning must — if it is to be done properly — be based on the earlier cognition (the results of positive research) and the solutions — designs — obtained must be compatible with socially accepted value criteria.

In the overwhelming majority of cases of prospective scientific procedure, due to the above-mentioned complexity of studied objects, the verification of the results of this procedure takes place not in the course of laboratory experimentation, but after material realization of the results. It brings about the need for vicarious experimentation, experimentation on a model, that is for the immanentization.

Another, no less important, and perhaps even more troublesome, reason for immanentization is the fact that the scientific prospective procedure is by its very nature a 'multi-faceted' procedure, synthesizing various aspects in which reality may be studied.

Due to the development of the characterized research procedure the inter- and multi-disciplinary studies have been undertaken, focused on the problem, besides the 'usual one-faceted' research (the monodisciplinary one according to the classical systematics of scientific disciplines).

Scientific procedure brought about the necessity for constructing a language which would make possible the description of multi-faceted reality in all its complexity and in the fullest possible manner, while at the same time allowing the multidisciplinary body of research workers to communicate among themselves — it is worth noting that the research procedure described had also some influence on the modification of the institutionalized forms of doing research work.

The above-mentioned situation resulted also in an increased interest in the methodological problems connected with the scientific prospective procedure and, remarkably, not only among specialists in methodology, but also among research workers of various specialities. This phenomenon found its expression in the emergence of numerous methodological conceptions — starting from the end of the XIXth century.

It seems that the most significant results of previous activities aimed at integrating the methodological output concerning the prospective research, should be seen in:

(1) The acceptance by the scientific community of the term 'a system'. The term, not the notion; for one may enumerate more than ten descriptions of a system (not all of them meeting the requirements of a definition).

Apart from the terminological difficulties, however, the idea of systems research is becoming more and more widely accepted, and this is pointed out by a Polish author, R. Kulikowski, in the following words: "Admittedly, the generally accepted definition of a system is not yet in existence; still, the notion in question has been found useful in the course of considering many problems, both theoretical and practical". [1]

(2) The dissemination of fundamental, most general principles of conceptualization in multi-facet research, viz.:

— constructing a model (of a system *in abstracto*), in which there are defined the properties and relations between those properties, i.e., the structure of a system, and the individuals (things) having those properties at their disposal;

— viewing a system as a whole (it must be noted that the latter is not a simple sum of system components called elements but the result of their integration);

— studying a system with a view to its function, i.e., in the teleological aspect.

This consists, therefore, in treating the studied object as a specifically separated whole, i.e., in such a way, as would make possible an analysis leading to decomposition into components and the synthesis of components (their composition) into a complementary whole. The real object (the system in *concreto*) identified or materially realized on the basis of the model (the system *in abstracto*) is an element of the same class as the model.

This very idea is expressed in the words of Kravets: "Out of the whole gamut of trends characterizing the modern cognition there are two which deserve special attention, for today they comprise almost all spheres of cognition and they will, presumably, have a decisive importance for the future development of science. We have in mind the probabilistic and systems-structural approach to the study of existing reality". [2]

(3) The necessity was recognized of undertaking studies on the

methodology of systems research; some such studies are already under way. This concerns above all: formal means of analysis and synthesis of systems, means eliminating or at least limiting the arbitrariness of decomposition and composition of studied systems; means of formulation of synthetical criteria of evaluation of the systems behaviour; means affording possibilities for quantitative description of systems structure and numerical determination of criterial quantities characterizing the studied systems and the processes used in their control.

Formulation of the problem of complex research on science

The formulation of the initial problem of the science of science research and the resulting operational definition of science will be preceded by a review of various meanings given to the term 'science'.

(1) Science — understood substantively — is the whole of the enunciations, i.e. knowledge.

(2) Science — understood substantively — is the body of people (and the apparatus) engaged in one of the spheres of human activity.

(3) Science — understoood functionally — is the whole of formal procedures leading from a set of data, axioms or assumptions to theorems. Science in this sense is called the Method (note: 'correct with regard to method' = 'scientific').

(4) Science — understood functionally is the whole of activities of the people (and the apparatus) forming the science in sense (2).

In each of the enumerated meanings science used to be and still is the object of research. Science in sense (1) is the object of apragmatic methodology of sciences, science in sense (3) is studied by pragmatic methodology of sciences, while science in sense (2) and (4) is studied also by sociology together with economics,psychology, etc. Thus the problems of science in each of the above-mentioned meanings used to have and still have their allotted place in the system of various disciplines.

However, it was found that the study of science (in its individual meanings) by individual disciplines, while satisfactory from the point of view of certain ascertainments — of the cognition of some aspects of science, is not satisfactory from the point of view of practical social activity. This activity calls for prospective research resulting in methodologically correct recommendations.

The need for undertaking the studies on science, seen as practice, originated in connection with the fact that among numerous problems of social life there began to appear more and more problems requiring a scientific solution (in sense (3)) and — more important still — with the fact that without such solutions the solution of the rest of the problems encountered in life would not be possible. In other words, science became the indispensable preparation for every human activity. Once an importance of science and — at least partially — its place in the life of the society has been noted, then, according to the praxiological principle of rationality, the organizers of social life were faced with a major problem:

What is to be done in connection with science, seen as the practice of social life, in order to increase the results obtained in social life, given the resources at the disposal of the society? It must be noted that neither the resources nor the results are here understood in their narrow sense, limited to those measurable in terms of monetary units.

The problem thus formulated belongs itself to the abovementioned category of problems of social life, requiring scientific solution. The complexity of this problem forces us to classify it as one of the most difficult ones among those waiting for solution. The complexity in question is not recognized by all and this must be accepted as the principal reason of so small a progress of the self-knowledge of science. And, in turn, the primary source of this reason lies in the language in which the specific problems of science of science have thus far been formulated. Thus the reason is of a methodological nature — it consists of the shortcomings of the language indispensable for an adequate description of complex phenomena.

It is virtually impossible not to quote here large fragments of an address of Sadovski during the Second Polish–Soviet Science of Science Symposium in Katowice:

"The specificity of the present stage of development of the research of science consists above all in: (1) treating the organization and management of science as problems of the first magnitude, connected in a special way with the practical utilization of the results of science and necessitating urgent solution; this attitude is attested to by the extremely rapid increase of the number of research workers, of outlays invested in science, and by the tendency toward lowering of the effectiveness of research work; (2) an awareness of the necessity for executing the theoretical analysis of science as a whole instead of considering particular aspects of research activity. Hence all the features of the present state of research on science of science and all the controversies centered around the subject: the well known priority given to the sociological approach to science of science; the anxiety to obtain — as quickly as possible — the quantitative indexes and then to use them in the practice of managing science; serious difficulties in the integral understanding of the character of science (...).

The construction of models of science of science is the essential and undoubtedly the most complex task in the present stage of development of the problems in question. Nowadays, the majority of research workers agree that the science of science cannot be built by mechanical summing of data, which are at the disposal of individual disciplines analyzing various facets of science (...). In fact, however, research still follows the well-worn groove of traditional scientific disciplines.

The state of development of the science of science is very similar, from the point of view of its methodological contents, to the situation existing in semiotics, heuristics and in other domains of interdisciplinary research. There appears a task of generalized, complex study of an object (a system of signs, creative activity, science as a whole, etc.), but comprehensive models of those objects (particularly — methodological means of constructing models of this kind) are lacking. As a result the elaboration of problems still remains within limits of fragmentary scientific disciplines (...).

One of possible ways of overcoming the above-mentioned difficulties encountered in the construction of the science of science lies, in our opinion, in the utilization of the ideas and methods of general systems theory in the science of science research (...). It will be our aim to distinguish two... extreme cases of the utilization of the general systems theory in the science of science: (1) for the purpose of an analysis of the structural interdepedence of the construction of science (of its complexity) and of other kinds of human activity, viewed in this case globally, without the existing divisions; (2) for the purpose of constructing theories of those individual facets of scientific activity which might be presented as systems..." [3]

The problem in question is not specific to the science as an object of study, but concerns any research on any social activity. It is for this reason that the general model of the problem of social activity is accepted as the starting point of the formulation of the problem of complex research on the science of science.

The model of the problem situation of a social activity, viewed as a closed system, contains two subsystems: subsystem A — the studied social activity, subsystem S — the remaining social activity (Figure 1). The enumerated subsystems are connected by the follow-

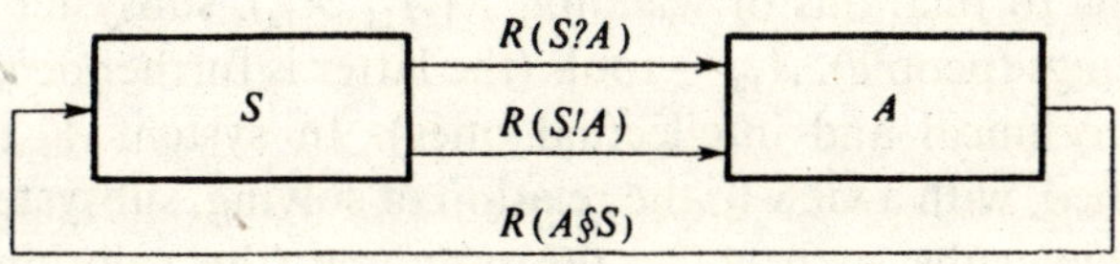

Fig. 1

ing relations: the essential relation $R(A?S)$ — the needs of subsystem S which are to be met by subsystem A; the control relation $R(A!S)$ — controlling subsystem A by subsystem S (defining the rigours limiting the degrees of freedom of subsystem A); the essential relation $R(A\S S)$ — the meeting of the needs of subsystem S by subsystem A in conditions determined by the control relation.

Considering A as a system (in this case S is an overall system) one distinguishes realizing subsystem A_1 and realized subsystem A_2 with a view to realization relation $R(A_1, A_2)$ — methods of realization procedure. This is shown in Figure 2a in the static mode and in Figure 2b in the dynamic mode.

Further decomposition of the model brings us to treating A_1

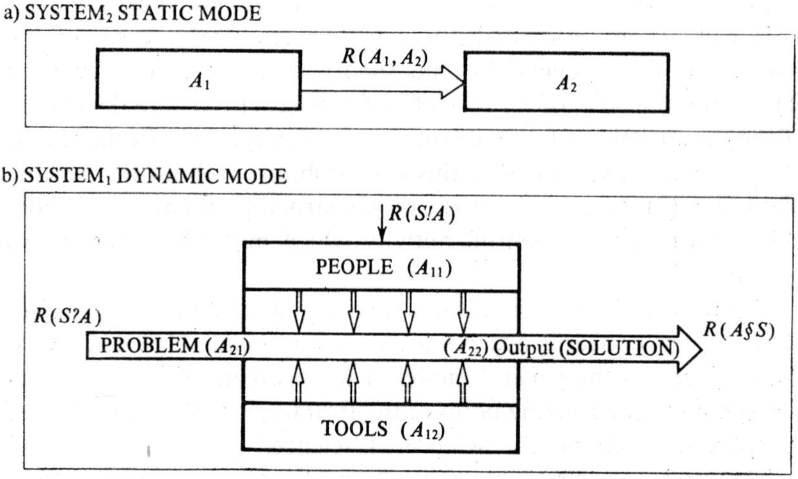

Fig. 2

and A_2 as systems, in which subsystems are distinguished with a view to relations defined (Figure 2b). In system A_1 there are distinguished, with a view to relations of wielding $R(A_{11}, A_{12})$, subsystems A_{11} — human beings (people), A_{12} — tools (the latter is further decomposed into the technical and intellectual ones). In system A_2 there are distinguished, with a view to the relation of solving, subsystem A_{21} — the problem, subsystem A_{22} — the outcome (the solution).

The model obtained as a result of the described decomposition is shown schematically in Figure 3.

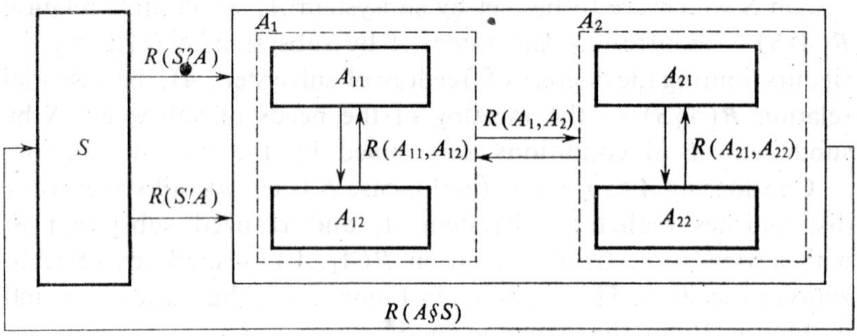

Fig. 3

The discussed model of the system of activity, interpreted in reference to science, is presented in Table 1.

TABLE 1

General model		Model interpreted in reference to science	
Name	Notation	Name	Notation
1	2	3	4
Overall system	S	Overall system	S
System of activity	A	Science	K
Essential relation of need	$R(S?A)$	Needs requiring scientific solutions	$R(S?K)$
Control relation	$R(S!A)$	Science policy	$R(S!K)$
Realizing subsystem	A_1	Scientists and tools used in research	K_1
Subsystem of people	A_{11}	Scientists	H_K
Subsystem of tools	A_{12}	Tools of science	T_K
Relation of wielding	$R(A_{11}, A_{12})$	Methods of using tools of science	$R(H_K, T_K)$
Realized subsystem	A_2	Research problems and their outcomes	K_2
Subsystem of problems	A_{21}	Research problems	P_K
Subsystem of solutions	A_{22}	Outcomes (solutions, theorems, recommendations) of research problems	O_K
Relation of solving problems	$R(A_{21}, A_{22})$	Methods of solving research problems (research techniques)	$R(P_K, O_K)$
Relation of realization	$R(A_1, A_2)$	Informational realization of research problems resulting from needs S	$R(K_1, K_2)$
Essential relation — meeting needs of S by A	$R(A\S S)$	Meeting informational needs of S by K	$R(K\S S)$

The problem situation interpreted in reference to science as a system of activity (*K*-system) is shown schematically in Figure 4.

In this way (by enumerating relations and subsystems of subsequent orders) the science (system) was described as an object of

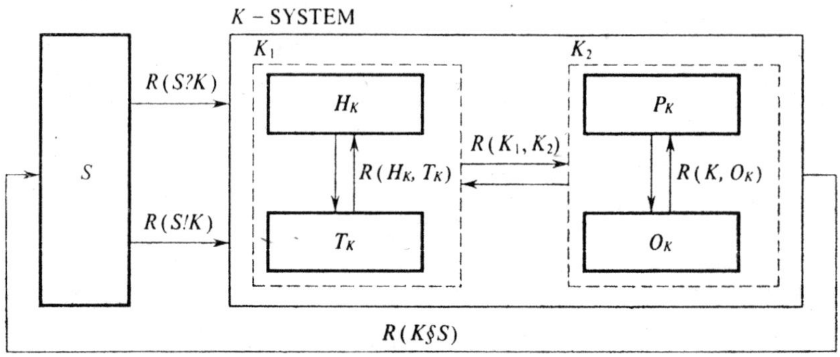

Fig. 4

complex research of the science of science. This description is derived from the previously given research problem, which may now be expressed in the following way:

What should be the structure of K-system and the relations between the latter and overall system S so as to assure their joint optimum functioning with regard to the criterion of organizational (systemic) effectiveness? It must be noted that the structure of the *K*-system signifies here the relations between the properties vested in the subsystems of *K*-system.

Problems of the science of science research — decomposition of first order of problem of complex research on science

While solving any given problem one follows the general procedure shown schematically in Figure 5 (after S. L. Optner). [4]

Taking into account that the solving of the initial problem of the science of science research in its entirety is not possible, one sees that it is indispensable to decompose this problem into subproblems.

The first order decomposition is not overly difficult, since the

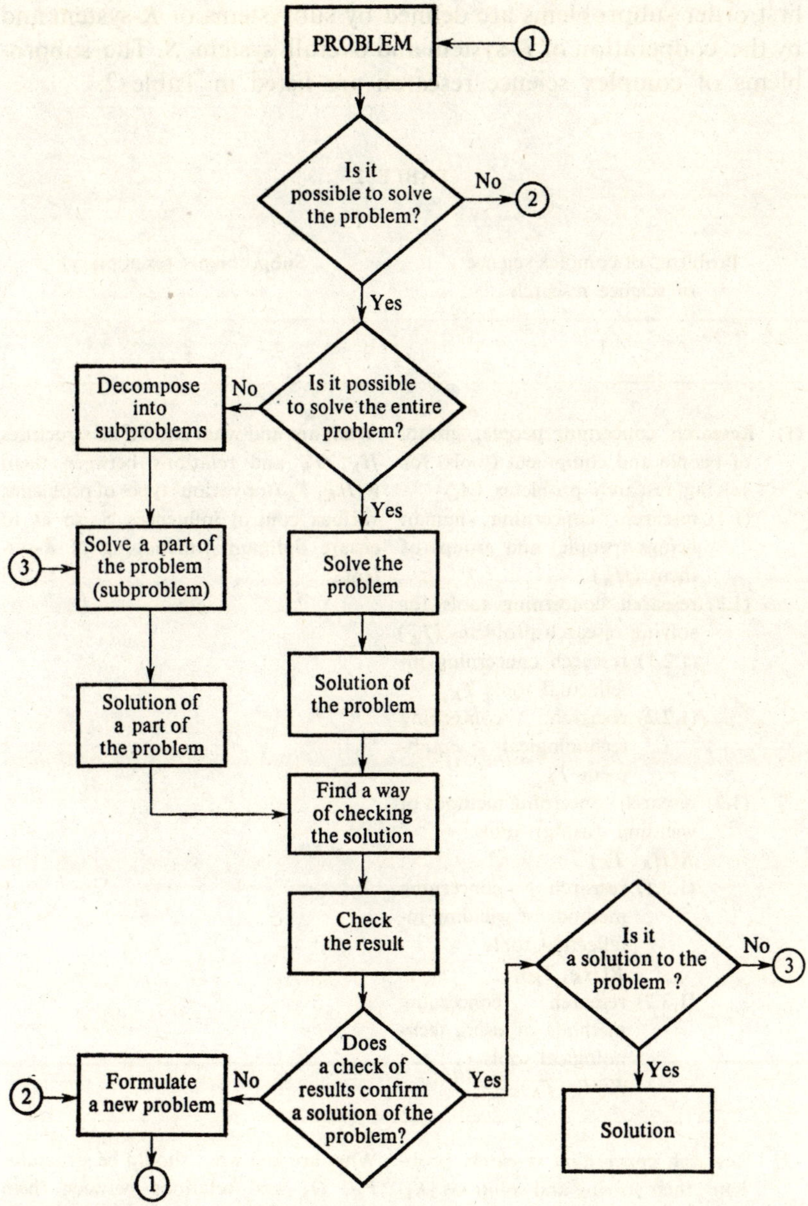
PROBLEM
1
Is it possible to solve the problem?
No
2
Yes
Is it possible to solve the entire problem?
No
Decompose into subproblems
Yes
3
Solve a part of the problem (subproblem)
Solve the problem
Solution of the problem
Solution of a part of the problem
Find a way of checking the solution
Check the result
Is it a solution to the problem ?
No
3
Yes
Solution
Does a check of results confirm a solution of the problem?
Yes
No
2
Formulate a new problem
1

Fig. 5

first order subproblems are defined by subsystems of K-system and by the cooperation of K-system and overall system S. The subproblems of complex science research are listed in Table 2.

TABLE 2

Problems of complex science of science research	Subproblems (exemplary)
1	2
(1) Research concerning people, groups of people and equipment (tools) for solving research problems (A_1) (1.1) research concerning human beings (people) and groups of them (H_K) (1.2) research concerning tools for solving research problems (T_K) (1.2.1) research concerning intellectual tools T_{K_1} (1.2.2) research concerning technological equipment T_{K_2} (1.3) research concerning methods of wielding (using) tools $R(H_K, T_K)$ (1.3.1) research concerning methods of wielding intellectual tools $R(H_K, T_K)_1$ (1.3.2) research concerning methods of using technological tools $R(H_K, T_K)_2$	What are and what should be structures H_K, T_K and relations between them $R(H_K, T_K)$ for various types of problems, various control influences S, so as to ensure optimum functioning of K-system? etc.
(2) Research concerning research problems, their solving and solutions (K_2) (2.1) research concerning research problems (P_K)	What are and what should be structures P_K, O_K and relations between them $R(P_K, O_K)$ for various needs S, various structures H_K and T_K, and various

TABLE 2 (cont.)

Problems of complex science of science research	Subproblems (exemplary)
1	2
(2.2) research concerning outcomes (solutions) of research problems (P_K)	control influences, so as to ensure optimum functioning of *K*-system? etc.
(2.3) research concerning methods of solving research problems $R(P_K, O_K)$	
(3) Research concerning research process $R(K_1, K_2)$	What conditions are and what should be for various influences of *S* on *K* and for various structures K_1 and K_2, so as to ensure optimum functioning of *K*-system? etc.
(4) Research concerning mutual influences of overal system *S* and *K*-system $R(S, K)$	In what way, for various structures K_1 and K_2, meeting of needs *S* is realized?
(4.0) Research concerning essential interrelation of *S* and *K*	In what way to foresee research problems resulting from future needs of *S* (forecasting)?
(4.1) Research concerning needs of *S* with a view to resulting research problems $R(S?K)$	In what way do solutions O_K meet needs *S*?
(4.2) Research concerning meeting needs *S* by solving research problems $R(K\S S)$	In what way are solutions O_K utilized by *S*?
(4.3) Research concerning control relation $R(S!K)$	In what way to shape science policy, so as to ensure optimum cooperation of *S* and *K*? etc.
(4.3.1) Research concerning control of *K*-system by social system	
(4.3.2) Research concerning control of *K*-system by remaining part of overall system *S*	

***Note*: Research listed in the table may be run in various temporal cross-sections (past, present, future) and on various scale (e.g., with a view to one type of problems, one scientist, one scientific institution, etc. — i.e., on the micro-scale, or with a wiew to all problems, the whole scientific potential, etc. — i.e., on the macro-scale). Research may also be run with a view to *K*-system and its cooperation with overall system *S* and also in the whole temporal range and jointly in the micro- and macro-scale. The last kinds mentioned are the truly complex ones.**

Decomposition of further orders — difficulties and ways of overcoming them

As may be seen in Table 2 the problems presented there (the exemplary subproblems) are still too complex to allow us to obtain a solution of any of them in its entirely. Therefore further decomposition is indispensable. The decomposition in question should take us to the formulation of research problems, which would be so detailed that their solution would be within reach of detailed methods ('research techniques') proper to various scientific disciplines (as viewed by the existing classification of science).

However, such decomposition is limited above all by:

(1) the shortcomings of the methodological knowledge;

(2) the limited character of the research potential available for science of science studies, above all as regards the qualifications and the disproportion in the structure of qualifications (a different number of research workers of different specialties);

(3) the limited financial means allotted to science of science research;

(4) the lack of an unequivocal criterion of organizational effectiveness, against which the functioning of *K*-system is evaluated.

For the above-mentioned reasons the detailed program of complex science of science research for a given moment of time depends on the state of the methodological knowledge and on the feasibility factors. Thus further decomposition of subproblems into subproblems of subsequent lower orders must be performed simultaneously with organizing and coordinating the course of research on *K*-system and its cooperation with oversystem *S*.

The necessity of formulating a comprehensive research program may, however, give birth to a danger of overwhelming the indications of the methodologically correct decomposition by the traditionally understood research of science and the interests of individual scientific units and institutions.

The results of all the studies on science, other than complex science of science research, should be treated as the fundamental information material, which will serve as a basis for individual research projects undertaken within the science of science. For it should be remembered that the studies on science will be performed not only by the science of science specialists of various disciplines (some of whom may in the course of this research become science of science specialists). For this reason the studies performed within various disciplines or on the basis specific to the latter are often and wrongly counted among the science of science studies, whereas they should be treated — from the point of view of the science of science — as various aspects (psychological, sociological, economic, historical, etc.) of the same subject. Hence the controversies which engage the rigorous classifiers of sciences disputing whether some studies of science are subordinate or superior to some others. For neither the history nor sociology nor psychology etc. of science are subordinate to the science of science nor is the latter subordinate to any of them. In each case the purpose of research, the point of view are different; the object of research is different too, though called by the same name of 'science'.

As far as the naming of a discipline as auxiliary in reference to another is concerned, we should first of all free the term 'auxiliary' from the pejorative connotation it is given here and there. The 'auxiliarity' (*sit venia verbo*) should be treated relatively: the specialist of science of science will aptly apply the term 'auxiliary' to the results of the research of a psychologist or of a historian of science, whereas the psychologist or the historian of science will apply the adjective in question to the science of science research.

Seeing that the complex research on science is of a multidisciplinary character, we must admit the fundamental importance of the coordinated effort of various specialists towards the cognizance of the essence of the matter in spite of the variety of languages of various disciplines. For this is the way in which science of science may become a discipline equipped with its own, distinct language. The way analogous to the way of other practical disciplines obtaining information which then makes possible the elaboration of practical models (design *sensu largo*) of man-made systems. (By a 'practical model' we understand here the pattern of a practical system or of a group of systems, allowing for and facilitating — indirectly or directly the realization of this system or of a group of systems, conceived with a view to the necessities of this realization.)

For this reason it is difficult to overrate the importance of the development of methodological research on systems of activity in general and research on science in particular.

Only the development of systems research methodology may contribute to the improvement of the correctness of this research, i.e., of the essentially proper decomposition of a problem into problems which could be solved by methods ('techniques') known to individual scientific disciplines, of their proper aggregation into the synthetic result, and of the proper evaluation (with a view to correctly formulated criteria) of the functioning of the *K*-system in the overall system.

The results of the science of science research thus obtained will make the possible the establishement of science of science as a distinct discipline.

NOTES

[1] R. Kulikowski, *Sterowanie w wielkich systemach* (*Large-Scale Systems Control*) WNT, Warszawa, 1970.

[2] A. S. Kravets, *Veroyatnost' i sistemy* (*Probabllity and Systems*), Izd. Woron Univ., Woronezh, 1970.

[3] W. N. Sadovski, 'Nauka o nauce i ogólna teoria systemów' ('Science of Science and General Systems Theory'), *Zagadnienia Naukoznawstwa*, No. 3 (15) (1968).

[4] S. L. Optner, *System Analysis for Business and Industrial Problem Solving*, Prentice Hall, Englewood Cliffs, 1965.

Ignacy Malecki

SCIENCE POLICY OF FOUNDAMENTAL RESEARCH*

Translated by Krystyna Kozłowska

Science policy as an integral part of politics

During the past fifteen years a fundamental evolution has taken place in the concept of science policy. Whereas in the middle of the Fifties the whole syndrome of problems centring round science policy was looked upon as having only a rather loose connection with the other spheres of social and economic life, today science policy is regarded all over the world, including the developing countries, as an integral part of politics. One reflection of this is that important government bodies, such as Ministries, Councils and Committees, are being set up to deal with questions of science, and that such bodies have a real impact on the decisions of other government bodies. It is not only individual governments that adopt a policy towards science. Groups of countries, too, acting together, are increasingly doing so. The Scientific and Technical Research Commission set up by CMEA (the Council for Mutual Economic Assistance) in 1963 undertook one of the first attempts in this field. It co-ordinated all the research being done by specialized commissions of CMEA. It also decided which were key problems for international co-operation, and gradually reduced their number, being thereby the first to bring about such a great concentration of research on the international level.

* First published in Polish as 'Polityka naukowa badań podstawowych' in Polish journal *Nauka Polska*, Vol. XX, 1972, No. 6.

Shortly after it was created in December 1960, the Organisation for Economic Cooperation and Development (OECD) — to which nineteen West European countries belong, as well as the USA, Canada and Japan — set up a committee for science policy. At the third meeting of the OECD countries' Ministers for Science, this committee issued a report on the organisation and development of fundamental research. [1]

In May 1972 the European Economic Community created a body which would coordinate science [2] in nine of its member countries.

We may also mention the role played by regional conferences of Ministers for Science and Technology in Asia (1968), Latin America (1966) and Europe (1970). [3] Nor should we, regarding the development of science in the western countries, omit to mention the part played by the science policies of great industrial firms, especially those with world-wide ramifications, such as IBM, Philips and Siemens.

In general, science policy embraces fundamental research, applied research, and development. According to the definition accepted by UNESCO, [4] this means every creative systematic activity undertaken to increase the body of scientific and technical knowledge, and to introduce new applications of such knowledge. The basis which should be taken for distinguishing research and development from any other activity is: "the absence, in the latter, of elements of originality or of (technical) innovation". Within the framework of science policy so conceived, we shall examine the question of science policy which deals with fundamental research. Obviously, just as science policy is closely bound up with social and economic policy, so must policy for foundamental research be treated in close conjunction with policy for applied research and development. Nevertheless there are certain characteristics specific to science policy, which we now propose to discuss.

Science policy consists of planning, coordinating, and directing scientific research at the general decision level. In particular it is the job of science policy to:

(1) fix the current and future priorities for research,

(2) distribute the resources needed for carrying out research,

(3) ensure the availability of research workers of adequate numbers and quality,

(4) organise research institutions,

(5) provide conditions conducive to creative research.

Here I should like to draw attention specially to the last of these points (which is frequently ignored in discussions about the tasks of science policy), for it is of fundamental significance. Objective conditions supported by administrative measures are not the sole determinant of scientific success. More essential still is the passionate dedication of the scientists themselves, for without it scientific research becomes merely a craft, bereft of those characteristics regarded as most essential to it.

Choice of scientific method is not the subject of science policy, nor is it the job of science policy to choose intermediate research goals leading to a final objective. But among the exceptions to this rule are cases where the purpose of political decisions is to decide between alternative methods proposed by the scientists; this may happen, for instance, in the case of great investment projects that involve the equipment of nuclear physics.

The scope and aims of policy with regard to fundamental research

The term 'fundamental research' is now universally recognised. It is used by politicians, civil servants, and scientists themselves, and it figures both in statistical reports and in articles in the daily press. In the literature dealing with the science of science, a good deal of space has been devoted to its definition. Formulae delineating the sphere of fundamental research are also to be found in publications issued by international organisations and in pronouncements by government bodies dealing with science policy in many countries. The Auger Report, [5] drawn up in 1958 at the request of UNESCO, on trends in science, had a great impact on later opinion. As a matter of fact the definition of fundamental research accepted in that report is still the one that is used in currently accepted statistical tables.

The OECD worked out for the Science Ministries of its member countries a number of definitions belonging to the sphere of science policy, and these were approved at the ministerial level at a conference in Frascati in 1966. [6] Hence the name 'Frascati definitions'. The American National Science Foundation, also, in 1967 submitted an official definition of fundamental research for the use of decision-makers in science, and this appeared in 1972 as a publication for the House of Representatives. [7] Similar definitions may be found in the Soviet literature. [8] Leaving aside certain semantic subtleties, these definitions obey two criteria which they have in common: (a) the aim of fundamental research science is to increase the store of human knowledge; (b) fundamental research does not directly serve non-scientific objectives.

Depending on how close the connection between fundamental research science and non-scientific aims is, we speak of 'pure' or 'free' research, carried on solely to increase the sum of human knowledge, and oriented research, which in the longer term leads to the attainment of non-scientific objectives. The more concrete the aim of fundamental research, and the closer that aim in time, the nearer is it in character to the character of applied research — hence the fluidity of the boundary line between the two kinds of science.

If often happens that those who order some piece of scientific research and those who carry it out are impelled by different motivation. Whereas the scientist is motivated mainly by the interest he feels in his problem, the 'client' who orders such research is mainly driven by non-scientific motives. This divergence has opened up the way to discussion on how far scientists themselves are responsible for the consequences of their discoveries, and in particular for the military applications of those discoveries. The Pugwash [9] conferences have provided a wealth of interesting material on this subject, which, however, is beyond our scope here.

What chiefly interests us in this paper is the motive behind scientific research, from the point of view of science policy. Briefly, we may distinguish four categories — four kinds in which fundamental research is linked to objectives laid down by science policy:

(1) 'pure' research undertaken to increase the sum of human knowledge;

(2) scientific research as an indispensible component of university at the undergraduate and postgraduate level;

(3) scientific research serving national, political, social and economic aims;

(4) scientific research constituing part of a 'vertical' project, from fundamental research to the application of scientific results to specified practical purposes.

Thus if we pass in succession from the first to the last of those categories, we find a closer and closer connection between scientific research and non-scientific aims, hence the categories differ somewhat from each other in their priorities.

Purely cognitive research

There is no doubt about it that the job of deciding whether or not 'pure' research is worth while belongs to the scientists themselves — that is, it is they who must weight up the chances of success when undertaking research in a given field. Figuratively speaking, the scientist should cast his net where he expects to catch fish.

On the other hand, pure research cannot remain outside the sphere of those who are interested in science policy. 'Big science' today calls for a great deal of money and for many highly qualified personnel. Moreover, it would seem only reasonable that the decision-makers, who provide the funds and the personnel, should have some say on the choice of subject and should be entitled to evaluate the results. Yet this matter is an exceptionally difficult and sensitive one for, according to the principles outlined above science should not be directed, but rather initiated and stimulated. When that is so, freedom of choice in research is particularly profitable, for it leads to new ideas and new discoveries.

In countries that are small or medium-sized it is not feasible to carry out equally intensive research in all scientific disciplines. Some scientific problems must be given priority. Here I have deliberately used the word 'problems', and not 'disciplines', for at the present state of scientific development any declaration about fostering the development of an entire scientific discipline would be too general to have any direct effect on the conduct of scientific research.

A number of factors may affect the choice of which problems should have priority. These are:

(a) *Predication of the possibility of new discoveries*, which as has already been remarked, is a *sine qua non* for undertaking scientific research.

(b) *The availability of what is needed for research*, and in particular creative, qualified scientists. Sometimes a decision has to be made between two extremes: whether to promote research on newly emerging problems, for which as a rule insufficient numbers of qualified staff are available, or to encourage vigorous existing 'schools' of science with ancient traditions and an international reputation.

(c) *International scientific cooperation*. Today international cooperation in fundamental research is much closer and more widespread than in applied research and development. This is partly because of the easier flow of information and the absence of restrictions such as patents, etc. For it is characteristic of the 'pure' research that results are published and thus available to all. Within the last forteen years, especially in Europe, many scientific programwes have been initiated and carried out jointly by several countries, mithin the framework of a single organisation, such as: the Low Temperatures and Strong Magnetic Fields Centre in Wrocław (Poland), the European Molecular Biology Organisation (EMBO) in Heidelberg, international centres for research in nuclear physics (Dubno-Sierpuchowo and Geneva). In each of the given countries the decision to participate in such research is usually a government one, which thus brings it into the sphere of science policy.

(d) *Coordination of scientific research in the sense of avoiding 'duplication'*. This aspect of science policy is sometimes put forward as the most important one. But probably not too much attention should be paid to it in the pure sciences. Of course, when expensive equipment is needed, duplication would mean a waste of resources, but on the other hand the same apparatus can sometimes be used by two teams working in parallel. The research being done on the microstructure of crystals at Grenoble University is an example of such an arrangement. In some other cases, there is nothing to be said against several research centres working simultaneously on similar problems. For since no-one can say beforehand what the results will be, each research centre individually may arrive at equally valuable results. Of course the available resources in staff and materials should not be spread too thin.

Apart from deciding on priorities, those responsible for science policy must also sometimes assess the work of the diverse research centres in order to be able to distribute resources. Despite all the attempts to make assessment quantitative, it cannot really be completely objective and impersonal. Efforts are made from time to time to discover criteria for comparison that will be valid on the international scale — for example, the number of scientific papers contributed by research centres to journals of international renown, or the number of scientists in such centres who have won national or international awards. [10] On the national scale, the most valid yardstick is the opinion of other scientists, and the bigger the circle of scientists, the more valid and objective this yardstick is. Yet scientists are only human, and their opinions can be tinged by subjective considerations. It is particularly difficult to arrive at a true assessment when only a small number of scientists are working on a given problem.

Hence in the sphere of science policy all assessment must be very cautious and based on as wide a range of opinions as possible. For the role of science policy is to nurture especially those research centres, or research teams, that are creative in their work, whose achievements are really new and of scientific value, and that have the courage to risk working on new problems.

Research linked with teaching

Today it would be a truism to say that university level teaching should go hand in hand with research. Basically this is the case with all branches of science, the only difference being the degree of closeness between the subject of research and the subject of teaching. Whereas in biochemistry, for example, this link may be very close, a mathematician who teaches the rules of the differential calculus, for instance, will do his research in the less classical branches of mathematics. Thus in the latter case it is not possible for the teacher to pass on his personal interests and scientific achievements to his students. Yet here, too, it is probably true to say that the professor's absorption in fundamental research will rub off onto his teaching, too, thus contributing to the flowering of a new generation of scientists and helping to make the given research centre a focal point from which new ideas radiate.

Two conclusions stem from this:

(a) Although it is possible that the resources allocated to research linked with university teaching are distributed in a 'thin layer' over the whole field of science, some differentiation does occur, depending on the rate of growth of the diverse disciplines, the amount of money and material resources needed for specific research, and the ability of the teaching staff to engage profitably in research. Such differences, however, do not stem clearly from the allocation of priorities mentioned above.

(b) The university structure tends to impose on all research that is linked with teaching the disciplinary rather than the problem oriented approach. Hence in many countries [11] scientific research that is inter-disciplinary in character, as well as research in those disciplines that have not yet found a place in the university structure, are to some extent treated as Cinderellas. This was the situation some time ago in biophysics and information theory.

This division of research according to university disciplines brings us to certain questions connected with the separation of fundamental research from other types of research namely: (a) is the boundary between pure and applied research the same as the boundary between the scientific disciplines? and (b) what is the place of the social sciences and humanities in fundamental research as a whole?

Many scientists would be inclined to answer the first question in the affirmative. But the matter is not a simple one, and the *a priori* imposition of schema in which certain disciplines are regarded as being at the top of the tree, and all research in them regarded as pure science, whereas it is assumed that other disciplines do no more than search for applications of already acquired knowledge, brings with it certain hidden dangers — a certain artificiality of division and the creation of privileges that are not always justified. We should amend the schematic division of disciplines into fundamental and applied, for two reasons:

(1) Within the last few decades mathematics, physics and biology have increasingly been becoming essential elements in many of the technical, agricultural and medical sciences. This evolution took place much earlier in chemistry; the distinction between pure chemistry and applied chemistry (seen in the name of the International Scientific Union) is today purely historical in character. Research on the electronic technology of semiconductors now lies right at the very core of modern physics, while human genetics and plant genetics are inseparable from a newly created 'basic' discipline: molecular biology. Institutions of advanced education such as the Lenin Polytechnical Institute in Leningrad, or the Technische Hochschule in Zürich, or MIT (Massachusetts Institute of Technology) carry out very advanced theoretical work in science in their Chairs or Technological Faculties, so we can easily be led astray by the name.

(2) The research that goes on in the theoretical sciences cannot always be said to be new or original. It is quite possible for the solution of a difficult mathematical problem by well-known methods to be of scientific value, and to qualify as a suitable subject for a Ph. D. thesis, although it may be doubted if it will 'increase the sum of human knowledge'.

Thus within a single discipline the research may be of different kinds. Pure science plays a bigger part in the 'theoretical' disciplines but occurs, too, in the 'applied' disciplines.

The place of the social sciences and the humanities in the total scheme of learning has been the subject of fierce discussion. In the socialist countries these branches of learning have from the beginning been treated as an integral part of Science. In the Anglo-Saxon countries it has been the custom for the term 'science' to embrace only the natural and exact sciences (with some additions, such as the 'behavioural sciences'), and for the social sciences and humanities to be given a separate place in statistics and organisation. But in the last five years in the Western countries, too, there has been a tendency to integrate the social sciences with the natural sciences, and to extend the term 'science' to include these two meanings.

Since, then, we may regard social and the humanistic sciences as part of Science as a whole, the question arises: What part of the research carried out in these sciences can be regarded as basic research? My impression is that here, it would be difficult to draw a sharp demarcation line dividing fundamental research from the rest. Surely we can say only that a given discipline has a preponderance of basic or of applied research. Much of the work done in the social sciences and the humanities consists of the very laborious collection of data (e.g., amassing material for a dictionary of dialects, or conducting sociological questionnaires), or field work (e.g. archaeological excavations). Although much of this work is purely technical in nature, it forms part of the programme of fundamental research, just as we may count as part of fundamental research the work that is done by the electron microscope in the science of molecular biology. Besides, the very work of collecting data may in itself call for originality — for instance through the application of new methods.

It should also be realised that technological and economic progress, which are being accompanied by a growth of population and increasing complexity of the social structure, does not diminish — nay, it even increases — the importance of the social and the humanistic disciplines. These advances have led to a situation where social science, which until recently was 'pure' — that is, theoretical — is (while continuing to have that character) now also becoming applied. The reason for this is as follows: present-day industrial technology (and, of late, biological engineering as well) is now creating hitherto quite unattainable possibilities for changing the conditions of life, and even man's biological make-up. Hence a tremendous social responsibility rests on those who must decide on the practical application of the results of the technical and biological sciences. Their decisions are all the more difficult in that far more alternative solutions are possible today. It would be nonsense to stop half-way and refuse to use the help of science, or to take advantage of the achievements of the technical sciences, or to pay no attention to the conclusions reached by the social sciences. This would be just same as making some complicated instrument with the help of science, but forgetting to say how it should be used.

The links between research and teaching are not confined to those at the undergraduate level. They are closer still at the postgraduate stage. In nearly all branches of science, progress now is so rapid that in the most dynamically advancing disciplines knowledge acquired at the university becomes outdated within a matter of a few years. Hence all over the world the accent is on refresher training for scientific staff. Indeed, in many fields of industry and medicine refresher courses have become regular exercises. Since the aim is to pass on information about the most recent scientific advances, the only person who can do this properly is one who himself stands in the front line of scientific progress. Anyone else undertaking this teaching job would have to pass on to his students someone else's results, taken from published papers, and so there would be a time-lag of at least a year. Another consideration is that such a teacher, not having himself done the work whose results he was passing on, would not be the best person to criticize and assess the worth of the teaching material.

In many countries, postgraduate education has become one of the most fruitful and durable links [12] between the universities and the people employed in industry, agriculture and medicine. This is an aspect which science policy as a whole cannot ignore — especially science policy dealing with fundamental research. The man who deals with practical problems is not so much in need of a recipe for what he is doing, as of a scientific understanding of the process he is directing. Sometimes hearing about a new discovery that has not yet been put into practice is sufficient to act as a stimulus to invention. The job of science policy is to encourage research in spheres that particularly interest people working at a practical job, for instance research on direct importance to practice. A lecture on the non-linear theory of elasticity may be very interesting and useful to the engineers working on the construction of some great dam, even if it does not tell them straight away how to reinforce concrete. Another task of postgraduate education is to make contacts between scientists and laymen in a particular field more efficient and more systematic. In some countries the entire organisation of postgraduate education comes under the aegis of the Ministry of Science and Technology, and so formally is treated as part of science policy.

Research connected with socio-economic objectives

It is this group of fundamental research that science policy is interested in particularly. When larger funds are allocated to fundamental research, this happens mainly because of this group and its objectives. Both the bodies responsible for the development of fundamental research and the scientists themselves are perfectly aware of this fact. In the years 1966–1968 the Council for Scientific Research in Great Britain, the Centre National de Recherches Scientifiques (CNRS) in France, and the Max Planck Gesellschaft in the Federal Republic of Germany issued formal declarations [13] stating that their job was not only to encourage science itself, but also to make it an instrument for the development of their countries. In the socialist countries, a conviction that science is of crucial importance to social and economic development lies at the basis of science planning and its key targets. [14] It has led to the setting up of Academies of Sciences. In both the Socialist and the Western countries leading scientists have made personal statements that science should serve the community, and declared they will help their governments to fulfil their economic plans. But universal agreement and willingness to cooperate are not sufficient. It is sometimes said that scientists tend to keep in line, to 'sail with the wind'. They do not wish to commit themselves. [15] This is part of the criticism increasingly levelled against science in the Western countries, especially the USA, from 1968 onwards. Without going as far as this view, and assuming that all the interested parties are acting in good faith, we may, as in previous sections ask: 'What are the objectives of science policy, and how are they to be achieved'?

In trying to answer this question, we come right at the start to a very controversial question: how much does fundamental research contribute to the socio-economic development of a given country? It is a fact that about 90% of the potential for fundamental research is concentrated in the twenty most industrialised countries of the world, that these countries account for about 80% of the world's industrial output, and that they are almost the sole initiators of technical innovations.[16] They are also in the best position to apply the results of science. But is economic growth dependent on the development of fundamental research, and if so to what extent? Attempts have been made to assess what percentage of growth of the national income is attributable to the practical application of technical innovation and the results of scientific research.[17] But these calculations are based on a number of arbitrary assumptions so can only be taken as estimating the size of the contribution made by science to the national economy. It is still more difficult to assess the part played by fundamental research. At one time widespread discussion was aroused by a book by Schmookler,[18] who studied the stimuli and starting-points of 934 inventions in agriculture, the paper industry, oil refining, and the railways, for the period 1800–1957. In his opinion, none of these inventions was directly inspired by a scientific discovery, for they stemmed either from a desire to solve some technical difficulty, or to win new markets. On the other hand in the great majority of cases the inventory, although they did not start off from some scientific discovery, had actually made use of such discoveries though frequently quite old ones made ten or twenty years previously.

The conclusions reached by another author, Trace,[19] will be of interest to us here. He studied how five inventions of great economic or social significance: magnetic ferrites, the Pill, magnetic recording for television, the electronic microscope, and the isolation of free radicals of chemical compounds, came to be made and applied. Analysing the course of discoveries and inventions dating from the 19th century onwards, Trace declared that about 70% of progress, has been due to 'pure' research, 20% to 'oriented research', and only 10% to development.

This estimate would seem to be somewhat exaggerated in favour of pure research, yet it is a fact that the farther back one goes to the root of technical innovations, the more important is the role of pure science. Those who believe there is a direct causal connection between a scientific discovery and advances in industry cite as an argument that the interval between a discovery and its practical application is frequently very short.[20]

It may seem paradoxical that it is easier to estimate the effect of basic science on the development of all the industrialised countries as a group, than its effect on an individual country. Apart from a few cases, the gap between a scientific discovery and its practical application is now of the order of 3–5 years. Between the making of a discovery and its appearance in print or its submission as a paper to a scientific congress, the average interval is a year. Now if a discovery has been made in country *X*, that country will have four years to see it applied, whereas country *Y* will have only three. But if country *Y* has a more rapid rate of technical progress, there is every chance that it will be able to catch up. This is of course a very primitive example, serving only to illustrate the fact that the application of scientific results knows no national barriers. We may recall the remark made a number of years ago by Lord Hailsham, then minister for science and technology in Great Britain, who complained in Parliament that although British scientists won Nobel Prizes, in actual fact their work was bearing fruit on the other side of the Atlantic, and thus increasing the gap between Great Britain and her richer partner.

It is also difficult to assess the effects of fundamental research because its results can be applied in very disparate branches of the national economy. Research on the plasticity of materials, for instance, is useful to the building, mining and engineering industries as well as to geology. So in assessment the right path should be to start with a given scientific discovery and trace it forward to its various applications, whereas until now assessment has generally started with a given practical innovation and worked back to see what part fundamental research has played in it.

Although quantitative assessment of the effect of basic research on socio-economic development is difficult, this does not mean that such an effect does not exist. It undoubtedly does, although owing to the lack of objective quantitative criteria both the public and the politicians are inclined to overestimate or underestimate the practical effects of fundamental research. The over-optimism about the profitability of fundamental research which was current in the

USA in 1962–1966 led in the end to a wave of disillusionment and pessimism, while President Johnson made acid remarks about the 'hard-headed' scientists.

Such fluctuations of mood show how important it is for those who decide science policy to be cautious when defining aims and means. Over-optimistic promises may achieve short-term targets, but in the end can lead to disappointments that are disastrous in their effects, particularly on the prestige of science.

When pessimism reigned, on the other hand, there was a slowing-down of the rate of growth of total expenditure on development and research in the big, highly industrialised, countries. (Table 1).

TABLE 1

Average per annum growth of expenditure on science (OECD data), percent

	1961–1967	1967–1970
France	18.7	7.4
Federal Republic of Germany	17.3	10.0
Great Britain	6.2	4.4
USA	8.4	4.4

On the other hand, pure science's share of the total expenditure on research and development has continued to grow (except in Great Britain). The yearly growth of budget expenditure on fundamental research has remained within the limits of 18–20% a year for France and the Federal Republic of Germany, and at 11% for the USA.

In Europe's other capitalist countries, the rate of growth varies within fairly wide limits, but in the highly industrialised countries it averages about 8%. No details are available for the socialist countries, but roughly speaking it may be concluded that the budgets for fundamental research there show a similar rate of increase. However, the factor which is decisive for science policy is not whether each year brings an increment in expenditure on science, be it big or small. The important thing is not so much the allotment of the yearly increments, but the structure of the total budget expenditure, which should take into account social and economic changes, as well as changes in the evolution of science itself. In other words, the objectives set before problem oriented research must always be kept up-to-date. It is more important that science policy should indicate the objective of oriented research than that we should try to arrive at quantitative global assessments.

Within the last ten years the objectives have altered noticeably, the changes having several features in common. Above all, economic aims and social aims are now much closer to each other than they were, both at the government and at the private company level.

In the socialist countries, this reorientation is now largely an accomplished fact, whereas in the Western countries it is no more than *in statu nascendi*, and today it would be hard to say how far the declarations made in the Brooks Report, [21] for instance, will be put into practice.

A country's economic growth or a enterprise's expansion naturally depend on a multitude of factors. But what we are interested in here is how the economy of a nation or the expansion of a business is determined by technical progress, which in turn stems from application of the achievements of pure science. In other words, the goal set before oriented fundamental research is technical progress which, it is hoped, will lead in the long term to economic progress. Here technical progress is understood in the wide sense, as including, for instance, improvements in crop-growing and stock-rearing. The chief economic results of technical progress are: reduction of the costs of production and exploitation, and improvement in the quality of products and services.

Thus science indirectly increases the economic effectiveness of: (a) industrial production, mining and the building industry; (b) transport and communications and (c) agriculture, forestry and fishing.

Research on the economy itself, and on management, occupies a place of its own. The social aims [22] towards which such research is mainly directed cover three principal fields: (a) improving the health of the population; (b) raising the educational and cultural level of the public; and (c) promoting social changes, e.g., interhuman relations.

But this scheme has an important gap in it: it does not take into account the social consequences of economic activity. It is to fill this gap that the aforesaid reorientation of objectives is taking place. Above all, high priority is now given to combating the deleterious effects of industrial growth. Man's environment must be protected — this is a problem which in the four years between the UN Secretary General's declaration in 1968 and the inter-governmental conference in Stockholm in 1972 attained its proper rank on the world scale.

In the Western countries, the main reason for not paying attention to the harmful social effects of industry is the fact that it costs a lot of money to diminish them. In the socialist countries, the fact that the protection of man's environment does not always get full attention is often due to organisational reasons — such as the parochial policies of the ministries or economic enterprises involved. A notorious case of this kind in the Soviet Union was the dispute between the Ministry responsible for the timber industry, and the Academy of Sciences, about the construction of a cellulose factory on the shores of Lake Baikal. It would be an over-simplification to see a gap only between the kind of science that is directed towards technical objectives, and the kind of science that is directed towards social goals. The public are beginning to be anxious, too about the bifurcation of trends in the biological and the social sciences. The revolutionary discoveries made in molecular biology and genetics are opening up hitherto undreamed of prospects for influencing the evolution of plants, animals, and man himself. It may be that genetic experiments will provide a greater threat to mankind than the pollution of water by industrial plants. Consequently it is particularly important for social aims and scientific aims to go hand in hand here.

The socio-economic aims facing science are not identical throughout the world. They vary from country to country, depending on many specific local factors, especially the size, and stage of economic and social development, of the country in question. But some rules common to them all can be perceived. For instance, although economy-oriented science ranks differently in importance in different countries, the problems it deals with are universally the same.

Conversely, the social aims towards which science is directed in many countries are fairly similar to each other, whereas the actual problems differ, depending both on the character of local research, and on the particular political and social circumstances. This rather abstract assertion may be illustrated by examples. Only in those countries that have a very advanced electrotechnical industry can research on the superconductivity of metals be undertaken. Genetic research connected with rice-growing is only of interest to some countries in certain geographical regions. Informatics on mass telephone communication over very long distances may certainly be very important, but only to big countries with a very extensive telecommunications network. On the other hand, centres doing research on superconductivity (either through the genetics of rice or through information theory) are all engaged in the same field of research; the results can for the most part be used in any country, not just the one where the rice plantation or the telecommunications network is. But with social targets the situation is different. No government would fail to aim, at least formally, at improving the health, education and culture of its citizens. The choice of priorities for research depends on differences that are not so much qualitative as quantitative. The reduction of infantile mortality, or the fight against cancer, are problems that face the whole of mankind, except that quantitatively the first is more important to the developing countries, whereas the second is more urgent to the highly advanced ones. The targets of the historical sciences or of linguistics can also be formulated in such a way that they are useful to every country, but obviously science policy in each country will tend to concentrate research on the history, culture and language of its own people. Thus such research cannot be universal in character, and in the nature of things it needs to be carried out on the spot. The system (so widely developed in the technical field) of adapting to one's own needs the results of fundamental research in other countries fails here. The same also partly holds true of educational research and sociological studies. It is well known how much damage was done by transfering the Oxford or Sorbonne patterns of education to the developing countries.

We have said nothing so far about research in economics, law, and the dynamically developing science of management. They occupy a separate place, for two reasons: (a) it is obvious, from their very nature, that their aims are long-term socio-economic ones; and (b) the connection between science policy and politics in the strict sense is especially close here.

In many countries this has practical consequences of an organisational nature, in that such research often comes not under the science ministries or committees, but under ministries or other bodies dealing with the general administration of the State. So research in economics is to some extent scattered throughout various branches of the national economy, and bound up with their research programmes, which makes central coordination less effective.

The orientation of objectives brings the biologist and the technician closer on the one hand, and draws the economist, lawyer and sociologist together on the other. This process may encourage the creation of inter-disciplinary teams, and indirectly, too, it may bring about changes in the traditional research methods in the social sciences. Hence the strong current of change now perceptible in science policy may help to do away with some drawbacks, such as the narrow restriction and isolation of certain lines of science pursued by isolated, individual scientists in highly specialised fields and disciplines. Thus there are very profound reasons, rooted not only in the logic of the development of science itself, but also in the social functions of science, for the integration of research.

Fundamental research as part of industry's research project

The classic research project consists of three parts: fundamental research, applied research, and developmental research. The fundamental research that comes into industry's general research project is directed towards a much more precise aim than the kind of research discussed in the previous section. While results are not taken as a foregone conclusion, a time limit for them is usually fixed for, right at the very start, when the whole project is being set up, the amount of time and money to be spent on the project are already settled. Big science programmes are carried out both by governments and by industrial firms. Some of the first projects to be established on a huge scale aimed at the improvement of armaments during World War II. It is a well-known fact that after the war, too, a large percentage of research in many countries, especially the USA (about 30%) was directed towards military goals. It is a matter of discussion just how much of this research had any significant effect on armaments technology. Military experts in the USA are inclined to take a pessimistic view of this; according to the Hinsight Report, [23] in twenty improvements that were of cardinal military importance, about 8% of the new ideas were contributed by scientific research, all the rest being attributed to technology. It is very possible, however, that in this report the author deliberately diminished the role of science.

Within the last twenty years the biggest research projects — apart from military ones — have been connected with the development of nuclear energy and with space research. From the point of view of a country's general science policy, the key question is: how much do science projects connected with military aims and space research contribute to the development of civil technology, and hence to the economic development of the country? In the mid-'sixties, government circles in the USA, when trying to get more funds for these projects, used these very arguments; and on the whole the public shared the view that military research and space research would bring with them progress in all fields of technology, e.g., in electronics. At that time this really was the case — e.g., the technology of integrated circuits, or of modeling simulating systems — while civilian air lines got jet engines that had been developed by the military aircraft industry. But now the situation has changed, as has been pointed out, for instance, by Professor Brooks, [24] Chairman of the National Academy of Science's Committee on Science Policy. In his opinion such a great degree of specialisation has been reached both in space technology and in the newest branches of military technology (e.g., anti-missiles protection systems) that their civilian application has become very limited, and the benefit is out of all proportion to the huge costs involved. In France, too, there has been criticism of certain intentions to take a bigger part in space research and in the development of nuclear research for military purposes.

Of course one could hardly call for the abolition of all research undertaken for the arm race and defence purposes; but in reality such research could be limited, as indeed has already happened in the case of space research and nuclear research connected with armaments. Possibilities are opening, up, at least theoretically, for new research programmes aimed more at people's welfare. This brings us back again to the reorientation of research objectives, this time with reference to concrete projects that are sometimes called 'second generation projects', in contrast to the 'first generation projects' isolated from social needs. The formulation of these projects is sometimes aided by comprehensive studies on social and economic development aimed at discovering how bottlenecks in these fields occur, and how they can be remedied. One such study on a world-wide scale was carried out by a group of scientists, economists and managers belonging to the 'Club of Rome', working along with an MIT group of futurologists headed by Forester. *The Limits of Growth*, [25] a book which was the first fruits of the Club of Rome's work, gained world-wide fame. This is not the place to discuss the theses put forward in that book and a subseqnent one, which, as a matter of fact, attracted a good deal of criticism. The fact is, however, that this problem — the limits of social growth — has become an extraordinarily vital one, and the solution we choose will have a marked effect on many economic decisions, and, indirectly, on science programmes. Undoubtedly there is a need for studies of this problem on the nation-wide or regional scale, for they alone can lead us to concrete conclusions. The technological projects aimed indirectly at overcoming the limitations of growth may be said to fall principally into the following three groups: (1) the fight against pollution, (2) improvement of the balance of energy-consumption, (3) the use of natural resources, and the recycling of waste material.

A number of such projects have already been undertaken both on the national and the international scale. To mention only a couple of examples, there is a project for reducing the pollution of the Rhine, in which five countries are taking part, and another example is the complex programme for reducing noise, already started in France.

The reorientation of research objectives, which is visible at government level, is also reflected to some extent in the research programmes of private enterprises. All the large motorcar firms, for instance, are above all expanding their research on the development of engines that will not give off harmful fumes, while in the chemical industry a great deal of work is being done to find methods of purifying factory effluents.

In many fields the centres of developmental research in industry are carrying out fundamental research that is part of the overall programme for improving production. The share of fundamental research in development as a whole varies within fairly wide limits, depending on the country and the branch of industry. This is illustrated by Table 2, which was compiled by the OECD [26] for several branches of industry. The data are for the year 1967 (USA 1966). There have been no radical changes of proportion within the last ten years.

TABLE 2

Expenditure on fundamental research as a percentage of some firms' expenditure on research and development

Country	All inter-viewed firms	Chemical industry, including pe-tro-chemistry and pharma-ceuticals	Electrical industry	Aircraft and missiles industry	Machinery production	Means of transport and com-munica-tion
	1	2	3	4	5	6
France	4.0	14.1	3.2	2.5	1.9	—
Federal Republic of Germany	5.1	9.8	1.3	11.1	1.3	9.1
Great Britain	3.4	7.5	3.4	0.8	1.9	1.3
Holland	31.0	39.5	—	—	23.6	23.6
Belgium	6.5	9.8	3.7	—	1.3	0.7
Sweden	0.7	0.5	—	—	0.1	0.1
Norway	3.6	4.5	2.2	—	3.0	1.1
Italy	3.0	9.6	0.3	—	—	0.4
Japan	10.2	14.1	0.6	—	3.7	6.7
Canada	4.7	8.3	0.7	1.1	0.4	0.3
USA	3.9	11.7	0.7	1.2	1.7	—

It will be seen from Table 2 that the chemical industry has the largest proportion of fundamental research, and the firms that pay the greatest attention to fundamental research are those that work mainly for export (Holland, Japan). In their science policy, the big companies differ from each other considerably as regards the freedom they give to their scientists. Some companies, when deciding on a field of research, allow their scientists to choose their own subjects, but fix a deadline (usually three years). Nowadays it is becoming more usual, however, for firms to set their teams of scientists long-term concrete tasks that are not necessarily connected with production.

Funds for fundamental research

One of the most pressing tasks of science policy is to allocate annual and long-term funds for scientific research. It is a very difficult matter to differentiate, from the total funds available for scientific research in the whole country, the amount that can be given to fundamental research. It is calculated on the basis of how much money is needed to keep up the scientific centres, or groups of such centres, which is easy to determine. But institutions engaged solely in fundamental research are exceptional. So the distribution of research centre funds between fundamental and applied research is done by guess work. On average it is taken that in 'academic' research centres fundamental research accounts for between 70 and 95% of the total, as compared with between 5 and 15% in industry's research centres. In the schools of university level, most of the money and time available go towards teaching, although these institutions also work on fundamental and applied research and development. It can be taken that on average about 10–20% of the total budget (including staff salaries) of universities of world-wide renown is spent on fundamental research in the biological and technical faculties. In the arts faculties, the proportion spent on pure research can only be assessed arbitrarily, for most of the expenditure goes on the salaries of the staff, whose work would be hard to allocate between teaching and research.

TABLE 3

Fundamental research share of the total expenditure (excluding investment) on scientific research carried out by the universities in a given group of disciplines (UNESCO data for 1969)

Country	Biological and technical sciences	Medicine	Agriculture	Social sciences and law
	1	2	3	4
Belgium	76.5	46.1	44.0	74.4
Czechoslovakia	all further education		—	80.0
France	89.0	33.3	8.6	89.6
Federal Republic of Germany	all further education		—	90.0
Greece	65.4	33.5	25.3	—
Hungary	65.7	58.3	15.4	80.0
Ireland	74.6	64.3	30.9	69.1
Italy	70.6	—	—	100.0
Holland	85.2	86.2	87.5	93.6
Norway	69.8	63.7	25.4	65.9

Expenditure of fundamental research is expressed in planning as a percentage of the total expenditure on research and development. According to the data available for 1969 (UNESCO, OECD), this percentage for some countries is as follows:

Belgium	32.7%	
Czechoslovakia	17.9%	
France	19.5%	
Greece	23.1%	
Hungary	14.6%	
Ireland	13.8%	
Italy	25.4%	(excluding expenditure on research and development in industry)
Holland	49.3%	(pure and applied together)
Norway	25.8%	
Poland	30.0%	(pure and applied together)

Romania	26.6% (pure and applied together)
Great Britain	10.5%
USA	14.0% (1966)
Federal Republic of Germany	23.5%

Statistics for the capitalist countries comprise only part of the developmental research carried on in private industry, so that in actual fact the percentage share of fundamental research is rather lower than the figure given.

In looking at total expenditure on science on a national scale, it should be remembered that in the capitalist countries some part of fundamental research is carried on by private firms. The OECD data of 1967 give us a picture of private industry's share in fundamental research:

France	13.1%	Italy	13.1%
Federal Republic of Germany	19.4%	Norway	8.5%
Great Britain	21.7%	Japan	23.9%
Holland	35.1%	Canada	8.7%
Belgium	21.2%	USA	19.5%

There is not much discrepancy in this table. As is only to be expected, the highest percentages are to be found in those countries (Holland, Japan) where industry is geared mainly for export. It will be observed that the ratio between fundamental research and total expenditure on research and development is not related to the countries' level of economic development. Those countries whose economy is backward (Greece) or whose industry is small (Norway) spend quite a lot on fundamental research partly because research in the universities should be encouraged, and partly because they are ambitious to do 'pure' research. Expenditure on fundamental research has a threshold below which such research is really fictional. In countries where experimental and developmental research is not much advanced, even such a small sum represents a rather large percentage of total expenditure. In the medium-sized industrial-agricultural countries, the ratio of fundamental research to total research and development varies between 12 and 20%.

But no theoretical recipes can be suggested as to how this figure is to be established. In many cases what is important is not so much the absolute figures as the rate of budget growth. Above all, additional funds in the budget mean the possibility that new staff can be engaged — that an influx of young blood and new ideas is possible. In the OECD publication referred to above, a minimum of 15% is recommended for the annual increase in the budget of the research institutions, as being the minimum required for them to develop properly. On the national scale, growth of the order of 10% of the current budget on fundamental research should probably be regarded as the lowest possible limit. Of course growth in the numbers of personnel in the research centres cannot go on without end, and of course there is always an infusion of 'new blood' through the rotation of older employees, but firstly it is a well-known fact that all over the world the mobility of scientific staff is very small, and secondly an increase of 10% more or less corresponds to the average growth of technical potential in the industrialised countries so the ratio between the development of science and the growth of the economy remains more or less the same.

The resources available for research depend on what the research is aimed at, on the amount of finance available, and the supply of scientific staff. Whereas current finance can be made available immediately if needed, and whereas between two and three years are needed for investments in buildings and equipment (apart from very 'big' apparatus such as astronomical telescopes), it takes at least five years to train a research scientist. One must also reckon with an interval of several years between a scientific discovery and its application, so only long-term or medium-term planning can deal with the cycle: fundamental research — practical application. It is only at this distance that we are capable of perceiving the non-scientific profit from fundamental research.

The view that scientific discoveries cannot be predicted, and that the planning of fundamental research is no more than a fiction, can still be quite often encountered. But probably it is more true to say that this negation of the value of planning is itself a fiction that is harmful to science. For, in the case of fundamental research, planning primarily commits those who order such research, and guarantees that the research will be able to proceed at a foreseen tempo, whereas the commitments resting on the scientists as a result of planning are much less precise, and to a large extent are moral in character.

It is the same with risk: the scientist stakes his reputation and his pride on the obtaining of worthwhile results, whereas the sponsor's stake is a financial one. Naturally the two sides are not comparable. The matter has been mentioned here only to stress the difference between the mechanism of financing fundamental research and its development.

The organisation of fundamental research

In nearly all the highly industrialised countries, scientific research is organised in three sectors: (1) the universities and other institutions of higher education, (2) the research institutes of 'academic' type, and (3) research centres connected directly with practice, e.g., industrial production.

As has already been stated, most of the fundamental research is carried out in the 'academic' institutions, and least in the institutions belonging to industry. Both in the socialist and in the western countries, many objections have been made to the way the institutions in all three sectors are organised. Within recent years the university structure has come in for most of this criticism. The university structure, it is said, is not a suitable one for the conducting of research, because it:

(a) is based on the traditional division into scientific disciplines, and so is not suitable for carrying on inter-disciplinary research;

(b) tends to lead to a 'spread' of funds, which in turn hampers the concentration of research;

(c) is very rigid; to change it would require many years, and generally would not lead to real alteration in the themes of research.

All these objections are nothing new. It will be recalled that in 1530 François I set up the Collège de France, in order to break the resistence of the Sorbonne, which refused to introduce the teaching of mathematics, Greek and Arabic. Examples can be found in our own times, too — for instance the English universities' hostility to research in genetics at the beginning of the present century.

There is no doubt that in the 'academic' institutes and in industry's research centres the work of an entire organisation can be focused more easily on some key projects, and there is no teaching to necessitate the demarcation of research into various separate disciplines. But rigid organisation is a principal obstacle here, too. Whereas in the universities there is some reason for maintaining the existing structure and range of subjects because of teaching needs, in those institutes which do no teaching, but only research, rigidity of research projects leads to complete bogging-down. As has been remarked by Weinberg, [27] this sometimes happens when an institute that has hitherto concentrated on certain problems, switches over its research to a chosen discipline, which can bring a research centre of the academic type immortality.

The directors of the Centre National de Recherche Scientifique (CNRS) in France reckon that from the moment of government decision to the full functioning of an experimental research institute, 6–7 years are needed, [28] but to do away with an institute whose work is no longer needed may take even longer. On the other hand, it must be admitted that some organisations are very bold in this respect. Between 1965 and 1970 the Max Planck Gesellschaft in Federal Republic of Germany, abolished or passed over to the universities thirty-one of its research centres (out of a total of eighty-one), while setting up a number of new ones. In Great Britain, the Medical Research Council between 1968 and 1970 established thirty-five new science laboratories, while at the same time doing away with twenty. It often happens that an institute, while keeping its old organisational structure, and its old name, becomes 'reborn' by going over to a completely new line of research. To take just one example, the Chemistry Institute in Mainz (belonging to the Max Planck Gesellschaft) has existed uninterruptedly since 1912, but within this time has changed its field of research entirely four times over. It began with the chemistry of food and the determination of molecular weights, then it went over to research in the chemistry of compounds of silica and boron. For twenty years after the last war it concentrated on nuclear chemistry. Recently the institute has been working almost exclusively on space chemistry and on mass spectroscopy — fields which are very far removed from classical chemistry. To take another example, the Institute de France, with its ancient traditions, now specialises in quite different fields than before. Today the work of its department of experimental medicine is confined almost exclusively to study of the fundamentals of immunology.

All administrative changes involve a serious danger that what has been achieved already will be lost altogether, and that research teams will be dispersed. So science policy concerning the organisation of research institutions should be extremely cautious. Rather than reorganising, better by far to make the organisational framework elastic enough to cope with a turnover in the themes of research.

In research centres connected with practical problems, provision for fundamental research is generally included in the overall projects for research. Only in the very biggest industrial organisms are 'futurological' laboratories independent. There the production and sales managers have no right to any say in the kind of scientific work that should be carried out by these laboratories. The objectives of scientific research are decided on by the top management in conjunction with the scientists (e.g., Farbenwerke Hoechst, IBM). When the research flourishes, and expands to cover a wider range of subjects, more staff are taken on, which leads sometimes to the subdivision of laboratories, or to the 'budding' of separate teams.

Table 4 shows the size of research centres in various countries (UNESCO data for 1969); the figures refer to research centres in the 'general services' sector, which includes all non-university research centres not directly connected with industry.

A closer analysis of the data shows that the very large institutes are in medicine, meteorology, and nuclear physics. Apart from a very few exceptions (e.g., the Pasteur Institute in Paris), those institutes which engage in fundamental research rarely have more than a hundred highly specialised staff. It is not easy to draw general conclusions, but at any rate one element of science policy is to decide on the optimum size of the research centres, although of course maintaining great elasticity.

TABLE 4

Country	Number of research centres employing the following number of scientists and engineers				
	less than 10	10–24	25–99	100–499	500 and more
Belgium	251	16	6	—	—
Czechoslovakia	21	27	104	12	—
Finland	54	11	7	1	—
France	72	26	18	15	4
Greece	73	9	1	1	—
Hungary	18	29	34	6	—
Ireland	30	5	3	2	—
Norway	149	13	3	1	—
Romania	9	8	47	7	1
Yugoslavia	60	35	28	4	—

The organisation of research in the universities and other institutions of higher education constitutes a problem in itself, which cannot be dealt with here. In the vast majority of countries, the authorities responsible for scientific and technical policy do not deal with the schools of university level, which in the USSR, for instance, come under a separate ministry, or in France under the Ministry of Education. Sensitive problems about the scope of responsibility then arise, for it is no easy matter to settle who is responsible for research, and who is responsible for teaching. In practice, a compromise is generally reached: the modest funds allocated for research that is connected with the teaching-matter comes from the university itself, while the finance for research projects that are not connected with teaching comes from outside sources. These extraneous funds are the real instrument of science policy.

It is probably true that personal contacts between scientists and industry are all-important. These contacts are sometimes more efficient even than the financing of university research by industry. One important factor is the mobility of staff in both directions. On the whole it is more common for staff to move from industry to the universities, than the reverse. For instance, about 6% of the teaching staff at the engineering, mathematics, physics and chemistry departments of the institutions of higher education in Federal Republic of Germany have come from industry.

In many countries the universities are richest in numbers of research scientists, but are comparatively poor in equipment. There is a natural tendency, then, for the existing scientific personnel to be made use of not only by the government, but also by industry. The university (or at least some of its faculties), is a meeting point where the science policy of the government, and the science policy of private industrial companies converge. In the Western countries these policies are not identical. One illustration of this statement is the famous highway 128 outside Boston, where a number of firms that use MIT results have occupied sites near that Institute. These firms have a big influence on the kind of research that is done in many of the MIT's departments. This instance is also an excellent illustration of how close links can be forged between a scientific institute and industry. These are somewhat analogous to the links between the Siberian Centre of the Academy of Sciences of the USSR (the Akademik Gorodok) with industrial laboratories and plants in the Novosibirsk region. In this latter case, however, the Siberian Centre's work has much wider repercussions, and covers a much wider range, including studies of volcanic activity on Kamchatka and the biology of Lake Baikal. The creation of similar or other ways of linking university and 'academic' research with the problems of practical life is one of the main tasks of science policy. Many other forms of such links can be examined with profit. In Britain, for instance, scientists provide consultations for industry (Cambridge Consultants Ltd). And in several of the West European countries there are state 'Research Agencies' which make the results of university research available to industry (e.g., ANVAR: Agence National de Valorisation de la Recherche, in France).

As has already been remarked both in the socialist and in the capitalist countries in recent years industry has been increasingly interested in permanent links with the universities with regard to postgraduate and refresher courses.

To sum up, the main problems of science policy as far as organisation of research is concerned are as follows:

(1) how to make existing structures more elastic, so as to keep the themes of research always up-to-date;

(2) how to support new research trends, particularly interdisciplinary ones;

(3) how to forge bonds between the research centres and practical life ensuring the stimulation of research and at the same time ensuring its development, while not allowing it to be shackled by the short term needs of industry.

The environment of research

Coming back to what was said at the beginning of this chapter, we must stress once more how important are the scientists' working conditions if research is to develop successfully. Hence the nature of these conditions must come within the scope of science policy. These conditions vary from country to country, according to local political systems, customs and traditions, and obviously can only be influenced by science policy to a limited extent. The most important factors in the scientist's environment would seem to be as follows: (a) financial position, (b) social status and authority, (c) encouragement and support in his work.

Often only the first of these factors attracts attention. It is certainly important, but is not the only one. At any rate, the scientist's salary should be such that he should not have to seek additional earnings, and such as to give him and his family a standard of living similar to that of people with the same level of qualifications in other professions. This is clearly the goal that the science administrators must aim at — whether or not it is attained will of course depend on the general wages policy of the government. Roughly speaking between 1960 and 1968 research workers' rises in salary were larger than the average rise for other highly qualified personnel.

As for social status, this problem is much more complicated than that of salaries, although in the Western countries (and particularly in the USA), level of earnings is the most important determinant of social status. The fundamentals of science policy differ qualitatively from one country to another, depending on the level of economic development of those countries. Among the greatest problems of the developing countries in this sphere are: the absence of a scientific community, and frequently the complete isolation of scientists from the ordinary social conditions in their country. These circumstances confront the science administrators in the developing countries with the difficult problems of how to maintain good inter-personal relationships, and how to prevent the 'brain drain'. Readers interested in this problem will find a wealth of information about it in the reports of international conferences.[29]

In the economically advanced countries, which have had a network of institutions of advanced education and research institutes for many years, the situation is completely different. The scientist is not just an individual, but also a member of a specific social group. According to sociological research,[30] the most distinct scientific community is to be found among the staff of the universities and similar institutions, and on the whole such staff enjoy the highest social status. On the other hand, research workers in industry, and to some extent in the institutes belonging to the Academies of Sciences, are usually much less known to the public, who are not really very aware of the importance of the work being done by these people.

Those research workers who are employed in scientific centres not belonging to the universities or the Academies of Sciences are also much more cut off from the scientific community in the course of their daily work. Of course administrative measures cannot do away with these discrepancies and social tensions. But it may be useful, for instance, to encourage scientific societies whose members are engaged in some special field, although their work places may be scattered.

Social prestige is not something that comes out of the blue. It does not arise through chance. It depends, for instance, on social usefulness. In the scientific world it depends on whether we can answer 'Yes' or 'No' to the question 'Is fundamental research of any use to the community?' In other words, is fundamental research needed. Such a need, if it does exist, not only enchances the scientist's social prestige, but also provides a direct stimulus to scientific research. Engels' well-known remark about need being the fulcrum of science is just as true today as it was then. Apart from all the possible distinction and awards that may come to a scientist from the State or from society, public recognition of the value of his work is not only his highest award, but also his constant stimulus. The scientist, apart from very few exceptions, cannot dream of the kind of fame that can come to writers or actors, but he is entitled to hope that the public will be well disposed in its attitude to science and scientists. For this reason, too, the government bodies responsible for science policy should study very carefully public criticism of science and scientists and act accordingly. The popularisation of science — one way in which science policy can exert an indirect influence — may play a beneficial role here.

We spoke in passing of the tensions that arise within the scientific community itself. This is a matter of exceptional importance to those working in fundamental research. On the one hand we have scientists working individually on theoretical problems, and on the other teams of scientists and technicians carrying out basic experimental research. Then again there are teams working on applied research and development. Not only is an atmosphere of mutual respect and understanding needed for cooperation and for the application of results, but it is also a *sine qua non* for the full development and flowering of creative ideas. Good relationships of this kind depend to some extent on the framework of organisation. Should fundamental research be carried on in separate institutions, or should we endeavour rather to bring both pure science and applied science under one roof?

Here we have merely drawn the reader's attention to the existence of some of the problems that arise in the scientific world. Deep analysis of these problems, changes in the organisation of science, and the informing of scientists as to how they can improve their own work, constitute still other aspects of those activities which we know under the common term: 'science policy'.

NOTES

[1] *OECD Communiqué A* (68) 16 and earlier publication by that Organisation: *La recherche fondamentale et la politique des gouvernements*, OECD, Paris, 1966.

[2] *Press Release*, May 1972, *European Community*, Brussels.

[3] *UNESCO Science Policy Studies and Documents*, Documents from various conferences.

[4] *Definition of Concepts used in UNESCO Surveys of Meetings of Governmental Experts on Science Policy*, Budapest, 1972, Dec. 3/6.

[5] P. Auger, *Tendences actualles de la recherche scientifique*, UNESCO, Paris, 1961.

[6] *La mesure des activités scientifiques et techniques*, OECD, Manuel de Frascati, 1966 (new edition being prepared).

[7] *National Science Foundation. Report of Activities*, Washington, 1968, and *Science Policy. A Working Glossary prepared for the Sub-Committee on Science, Research and Development of the Committee on Science and Astronautics US House of Representatives*, Washington, 1972.

[8] G. Dobrov, *Nauka o nauke*, Kiev, 1968.

[9] *Proceedings of the XVI Pugwash Conference*, Ronneby, 1967.

[10] J. J. Salomon, *Science et politique*, Du Seuil, Paris, 1970.

[11] J. Ben David, *La Recherche fondamentale et les universités*, OECD, Paris, 1968.

[12] S. A. Lakoff, *Knowledge and Power*, Free Press, New York, 1965.

[13] *CNRS. Rapport de conjouncture*, Paris, 1968; *Council for Scientific Policy. Second Report*, 1967, *Max Planck Gesellschaft Berichte*, 1966.

[14] See J. Kaczmarek, *Nauka Polska*, 1972.

[15] *Le système de la Recherche*, Volume I, OECD, Paris, 1972.

[16] *Condition du succes de l'innovation technologique*, OECD, Paris, 1971.

[17] *Economic Commission of Europe: The V Meeting of Senior Economic Advisers*, Red. Doc., Geneva, 1968; Y. Shcherbakov, *Nauchnaya, politika*. Institut Gornogo Diella Novosibirsk, 1970.

[18] J. Schmookler, *Invention and Economic Growth*, Harvard University Press, Cambridge, Mass., 1966.

[19] H. Trace, *Technology in Retrospect and Critical Events in Science*, Illinois Institute of Technology, NSF, Washington, 1969; see also E. F. Denison, *The Sources of Economic Growth in the U.S. and the Alternatives before Us*, Committee for Economic Development, New York, 1962.

[20] I. Malecki, *La physique dans la technique contemporaine. Melange*, Université de Bruxelles, 1977, p. 291.

[21] *Brooks Report. OECD Conference of Ministers of Science and Technology*, Paris, 1972.

[22] Z. Kowalewski gives an analysis of science policy in: *Nauki społeczne a rozwój społeczny* (*The social sciences and social development*), Pax, Warszawa, 1971.

[23] *Project Hinsight. Final Report*, Office of the Director of Defense Research, Washington, 1959.

[24] Press interview by Professor Brooks, Cambridge, Mass., 1972.

[25] *The Limits of Growth*, Club of Rome, Rome, 1972.

[26] *Le système de la Recherche*, Volume I, OECD, Paris, 1972.

[27] A. M. Weinberg, *Reflctions on Big Science*, MIT Press, Cambridge, Mass., 1967.

[28] *CNRS Rapport de conjoncture*, Paris, 1968, 1970.

[29] *Science Policy Studies and Documents*, UNESCO, 1970.

[30] *The Scientific Community. Analytical Study*, UNESCO, 1972.

Klemens Szaniawski

LOGICAL ASPECTS OF OPTIMIZATION*

Translated by Olgierd Wojtasiewicz

The literature concerned with optimization as a rule refers to special classes of problems and to techniques for their solution. The present paper, by contrast, will be concerned with what is common to the various problems of optimization, without regard to the specific issues resulting from additional assumptions.

This is, of course, not a new problem. This is why a considerable part of the paper will be a brief summary of what is otherwise known. The treatment of this subject can be justified by the fact that the terms *optimization* and *optimal (optimum)* are used very loosely, which is a bad custom, because those terms are evaluative in character, and hence predicating them about something often conditions specified actions.

Optimization, in its primary and the only unquestionable sense of the word, means answering the question: which action in a given set of actions is the best, i.e., maximizes a given linear order interpreted as a preference.

As can be seen from the above definition, the optimization problem is defined, and optimization itself is essentially performable (i.e., if we disregard technical problems, to be discussed later) if the following three conditions are satisfied:

---

* First published in Polish as 'Logiczne aspekty optymalizacji' in an internal bulletin of Warsaw University *Biuletyn Naukowy Instytutu Nauk Ekonomicznych Uniwersytetu Warszawskiego* (1974), No. 32.

(1) the set A in which maximization is performed is defined,
(2) the linear order $\leqslant$, of which A is the field, is defined,
(3) that order has a maximal element, i.e.,
$\bigvee_{a^* \in A} \bigwedge_{a \in A} (a \leqslant a^*)$.

An optimization problem is thus a pair $\langle A, \leqslant \rangle$, which satisfies (3). That condition is the so-called positive assumption of the question about a maximal element, which guarantees that there is at least one answer to that question. The logic of questions usually requires that the negative assumption, which states that there is at most one answer, be satisfied, too, but we shall not require that. If, as often occurs, more than one element of A maximizes $\leqslant$, then the set of all such elements will be treated as the solution. Actions or operations which are elements of that set will be termed optimal. And it must be noted that we shall often be satisfied when one such element is indicated, and we do not call for the complete solution.

The simplicity of this set of concepts, which contrasts with the complications which usually mark real optimization problems, can be explained by the general nature of the concept of action used here. The set A can have as its elements entities with any degree of complexity, e.g., strategies (in the sense imparted to that word by the theory of games: that is, rules each of which determines a sequence of reaction to circumstances which change in time and are not known in advance, but belong to a fixed repertory). This makes it possible to apply the same concepts to static and dynamic problems.

Note, however, that optimization defined in this way has fairly limited extension relative to the current use of that word. For instance — from the point of view adopted here — it is meaningless to speak about optimality of some action without relativization to the set in which optimization is performed. And such statements are quite prevalent.

As can be seen, optimization in the fundamental sense of the term is identified with decision making under certainty. It is a model which assumes complete knowledge and self-knowledge on the part of the decision maker in a given situation. A possible weakening of this assumption will be discussed later.

The form of (2) and (3) is logically the simplest one, but not the one most often used in practice. Usually, instead of an order of preference we have to deal with valuation using real numbers, called utilities. This is a stronger assumption, because numerical valuation generates a linear order (by the arithmetical relation 'not greater than'), but not *vice versa*, unless we add additional postulates. Thus instead of (2) and (3) we have:

(2′) there is a real function u defined on the set A,

(3′) the function u has a maximum, i.e.,

$$\bigvee_{a^* \in A} \bigwedge_{a \in A} (u(a) \leqslant u(a^*)).$$

The optimization problem then takes on the form of the pair $\langle A, u \rangle$.

Concerning the manner in which the numbers $u(a)$ represent the valuation of the elements of A it is usually assumed that it is defined up to the unit and up to the origin of the system (the so-called interval scale), which is to say that

(4) only those statements about the numbers $u(a)$ are meaningful whose logical value is invariant with respect to positive linear transformation.

From the point of view of optimization as such, condition (4) is irrelevant. The results of such an optimization would admit a much broader class of transformations, namely all monotonic transformations, i.e., those which preserve the linear order. Condition (4) is needed only in the case of optimization in the broader sense — i.e., because it enables us to compare expected utilities in a sensible manner.

But the question arises, why in practice do we not use the weaker variation of the optimization problem, i.e., $\langle A, \leqslant \rangle$, if its strengthening to $\langle A, u \rangle$ is not necessary? It is probably so because, if A has many elements (which is usually the case), we do not see any natural way in which the relation of preference could be given in that set, unless preference is fixed according to a numerical characteristic of actions or their effects (e.g., in terms of money). But in such a case we in fact mean the maximization of that characteristic, which we thereby assume to be a valuation. It is another question whether such a valuation satisfies (4). We shall not discuss the difficult problems which arise when utility is approximated by a magnitude which has not the properties of utility, e.g., by money. [1]

Thus practical considerations make us reverse the logical order: preference comes to be generated by utility although the latter, being a numerical representation of preference, is secondary to it. This is revealed in the theory which introduced the concept of preference axiomatically. The essence of the universally accepted idea of von Neumann and Morgenstern consists in that, having assumed a preference on a set A (the latter expanded to a set of lotteries in which the elements of A are prizes), we prove the existence of a utility function, defined on A, which satisfies (4) and is in agreement with that preference — subject to the condition that the latter satisfies certain postulates. [2]

So much for the concept of the problem of optimization and its formulation. Its solution is interpreted as the indication of an element of A which is maximal in the sense of $\leqslant$ or u, and the proof that it is really a maximal element. Both are achieved if the operations chosen to solve the problem have been designated by a method which has been proved to generate maximal elements.

Optimization which is essentially performable, i.e., given that conditions (1), (2′), (3′) are met, may prove impossible to perform for various reasons, of which we shall mention two, which seem the most important ones.

First, the elements of A may be so complicated that they do not lend themselves to identification, such as, for instance, strategies in a sufficiently complex many-move game, say in chess. In such a case we are not in a position to indicate the best strategy, for we are not in a position to indicate any strategy (even though the set of all strategies in chess is defined correctly). Note that, contrary to appearances, this is not a difficulty that could be removed owing to advance in our knowledge: the information required to designate so complex an entity as strategy in chess reaches dimensions which are beyond the capacity of information carriers in our universe.

Second, it may occur that the elements of A are simple entities (so that we can easily indicate them) and their comparison is not difficult either, for we can easily find the values of u associated with them, but there are so many of them that finding the maximal element by simple enumeration exceeds our actual capability to calculate them. This applies, for instance, to the well known 'problem of the travelling salesman' if the number of the towns is sufficiently large. [3] This kind of technical non-performability of optimization is not essential in nature: it can be removed by finding a method which replaces complete enumeration and enables us to solve the problem in a fairly small number of steps.

The last-named circumstance accounts for the fact that optimization in its strict sense gives rise to non-trivial problems, some of them even very difficult, in spite of the simplicity of the problem in its general formulation. The point is to devise a method that would make problems in a certain class technically resolvable (and that in a possibly simple way). Such problems were earlier either not resolvable (in the second sense of the word) or their resolution would require too much time and too large a resource.

We shall confine ourselves to these comments concerning optimization in the strict sense of the term. A broader, and practically rather metaphorical, interpretation of that term is arrived at when we weaken those conditions [4] which define the essential performability of optimization, i.e., (1) and (2) or (2′).

Condition (1) makes optimization relative to the set A. Like any set, the latter can be defined in two ways.

The first consists in enumerating its elements: $A = \{a_1, a_2, ..., a_m\}$, which, by the way, assumes that we are dealing with a finite set, and even with one which is practically moderately numerous. This way of defining A identifies it unambiguously. The result may be at most not correct relative to the purpose which such a construction is to serve (for instance, because an operation which should have been considered on account of its value was not covered by that enumeration), but leaves no doubt as to the scope of the set being defined.

The second consists in indicating a necessary and sufficient condition of membership in A: $A = \{a: a \in Z \wedge F(a)\}$, where Z stands for a certain superset (e.g., the universe of the discourse), and F is a property which designates the elements of A. [5] For instance, in linear programming Z is the set of all real vectors of n components each, and F stands for the conjunction of constraints.

The latter case, probably the more frequent of the two, can cause trouble due to the general terms used in defining A. If they are vague, then that vagueness (and hence the possibility of different interpretations) burdens the optimization problem, and hence the concept of optimal action. The trouble is real, but it is otherwise cmmon, because we have to do with the vagueness of concepts in every field.

The following case is more specific: Z is a very large set (e. g., the set of all possible actions), and the constraints F are such that they enable us to state *ex post facto* whether a given element of Z is in A or not, but they do not give the method of generating the elements of A. A typical example of such a constraint is the requirement that the operation in question should lead to the attainment of a certain goal (the assessment of operations or actions by the function u depends then upon the outlay of means and possibly also on some side effects).

The problem of finding the best operation is, under such circumstances, an open question. [6] The resolution of such a problem requires inventiveness, because there is no method which would reveal the operations to be considered. This is particularly difficult when the causal conditionings of the state of things postulated as the goal are complex and little known (which is a typical situation in the case of social phenomena). To acquire a better knowledge of such conditionings we usually carry out particular research, which, however, does not always prove sufficient to indicate some elements of A.

May we speak about optimization when it comes to such a problem? Strictly speaking, we cannot, because all the elements of A not being indicated precisely we are not in a position satisfactorily to define the function u, and *a fortiori* to prove that it has a maximum (which is not the same as its having the upper bound, the existence of which we can readily accept). This means that the conditions of essential performability are not met. Further, no tentative solution can be shown to be really a solution: there is always a possibility that better information or greater ingenuity can lead to our finding in A an operation which is better for the purposes under consideration. Hence, if we use in such a case the term *optimization* then we do not do so in the literal sense of the word, or we relativize it to a certain strictly defined subset of actions which make us attain our goal, i.e., actions which are in A.

We shall now consider the principal methods of weakening condition (2′). Formally they reduce to the replacement of a single utility function u by a family of such functions. For convenience we assume that their number is finite, namely n, even though in fact they may form a continuous set. We are thus now concerned with a set of n optimization problems:

(5) $\langle A, u_1\rangle, \langle A, u_2\rangle, \ldots, \langle A, u_n\rangle$,

whose common feature is the occurrence in them of one and the same set A of actions.

The family (5) of problems has two commonly known interpretations. The first is that the utility functions in question are to be interpreted as evaluations of actions under varying assumptions concerning the conditions under which a given action would take place. Hence a function u_i evaluates actions under the assumption that the i-th set of conditions has materialized. In the other interpretation the utility functions are treated as evaluations of those actions in various respects, so that a function u_i evaluates actions in the i-th respect. There is a very fundamental difference between these two interpretations, to be discussed below in greater detail. A few examples first.

A typical example of the first interpretation is offered by the case when actions are assessed in the light of their effects, and these in turn depend, for every action, on which of the n sets of conditions has materialized. We may assume in this connection that the principle of the assessment of those effects is the same each time.

The second interpretation, on the contrary, does not postulate changes in the conditions (and the effects of actions as well), but introduces n different criteria of assessment of one and the same action, e.g., the assessment of the various types of investment by the expected returns, the amount of outlays, damage to the environment, impact upon the social structure of the region, etc. This is often called multidimensional evaluation of actions in A.

It can easily be seen that the first interpretation of the set (5) of optimization problems is known in the literature of the subject as 'decision making under uncertainty', and the second, as 'decision making under plurality of criteria of assessment'.

In both cases the word *decision* is used here, and we may think that it is used not without reason, for we speak about decision when the choice of conduct is essentially a choice, i.e., is not predetermined. A decision should leave a certain margin of freedom to the person who makes it. In the case of optimization in the strict sense of the word nothing like that takes place, because the solution is unambiguously determined by the conditions resulting from the problem, the only point being that that solution must be found. It depends (or rather does not depend) upon the person involved in the same way as the solution of a correctly formulated mathematical problem. The preferences which occur in the optimization problem do depend upon the person concerned, but once they have been settled (and it is only then that the optimization problem becomes defined) they determine the choice of action without any further interference on the part of that person.

On the contrary, when choosing an action from the set A in a situation formally defined by (5) such a person does not face any problem that is unambiguously defined. He may accordingly be called a decision maker in the strict sense of the word, just because he makes a choice which is not merely apparent.

This is so, of course, except for one very special case, namely when there is in A an action which settles all the n optimization problems at the same time. It is then said to dominate all other actions in (5); the choice of that action, if it exists, has of course nothing arbitrary in it. Yet it would be unrealistic to postulate the existence of a dominant action.

Having pointed to the inevitable arbitrariness of the choice of action in (5) we have to recall that the theoretical work, known from the comprehensive literature of the subject, tended to limit that arbitrariness in some rational way. This might be formulated thus. The decision maker who strives to choose a course of action proceeds in two steps. The first consists in transforming the set (5) of problems into an optimization problem in the strict sense of the term. He does so by making use of a transformation which we shall denote by K:

(6) $K(\langle A, u_1\rangle, \langle A, u_2\rangle, \ldots, \langle A, u_n\rangle) = \langle A, u\rangle$.

The transformation K will be called the criterion of decision making or just briefly the criterion (that word being, of course, used here in a sense different from that when we speak about a criterion of assessment). The summary utility function u obviously need not be identical with any function u_i, and usually is not.

The second step consists in optimization in $\langle A, u\rangle$. Since the latter, is, in principle at least, a trivial issue, the point of gravity shifts to (6). Hence the limitation of the freedom of the decision maker finds expression in the conditions imposed upon K. It is these conditions which make the difference between the two interpretations of the set (5) of problems, for something else may rationally be expected from a choice made under uncertainty than from the choice made under the plurality of criteria of assessment.

Much attention has been paid to the axiomatic description of K for the first of these two interpretations. On the whole, the literature of the subject concerned with that issue is much richer. We shall not, of course, list again all those postulates, whose formulation and discussion can be found elsewhere. [7] Their number is excessive, in the sense that all of them cannot be satisfied at the same time (for they contradict one another). One must, therefore, choose from among them, which agrees with the claim that interference on the part of the decision maker is a necessary condition of transforming the set (5) into an optimization problem *sensu stricto*.

The postulates are usually formulated as the condition imposed directly upon the optimum set yielded by the application of the criterion K and optimization. We have separated here these two steps exactly in order to reveal their connections with optimization in the narrower sense of the term, and also because the criteria actually used take on the form of (6). Thus, for instance, the maximin criterion assigns to (5) such an optimization problem in which the function u is defined thus:

$$(7)\ u(a) = \min_i u_i(a).$$

For Savage's criterion [8] that formula takes on the form:

$$(8)\ u(a) = \min_i (u_i(a) - \max_a u_i(a)),$$

etc. If we admit probability distribution over the set of possible conditions of action (the question arises then, what that probability is and how we know its numerical values), then the function u obviously becomes the expected utility u_i; that case is usually called decision making under risk.

The postulates concerning K under uncertainty are *not* transferred automatically to the case of multidimensional valuation. The requirements concerning K are those which differentiate formally interpretations of the complex decision problem (5). Let us try to say, without going into technical details, in what that difference primarily consists.

Now, in accordance with the first interpretation the optimization problems in (5) form an exclusive disjunction. The decision maker has to solve one problem only, but he does not know which problem it is. Hence the fact that when making his decision he takes into consideration all the functions u_i is due to his lack of knowledge as to which set of conditions really takes place. Complete or partial elimination of the decision maker's ignorance (through providing him with appropriate information) thus improves the conditions under which he has to make his choice. In the limiting case, when information is categorical, exact and true (i.e., points to the real set of the conditions under which he has to act), it eliminates all the problems covered by (5) except one, and the decision maker can optimize his choice in the strict sense of the term.

Generally, in the case of the first interpretation, it is appropriate to speak about information supplied to the decision maker, and it is true to say that such information enables him to improve his choice (and hence has a certain value for the decision maker).

Such information approximates the problem of decision making under uncertainty to the optimization problem *sensu stricto*.

This can, of course, be formulated more precisely. One can also discuss problems connected with the typology of information and its use and valuation. [9] (It is worth noting that Wald's theory of statistical decision functions is nothing other than an analysis of the methods of using information of a certain type when making decisions under uncertainty). Without going into that, let us emphasize that an essential requirement concerning K is that the value — in the sense of K — of the information about the conditions of action should, on the whole, be a non-zero one (on the whole, i.e., at least when that information is categorical, exact, and true).

Evaluation of actions in many respects simultaneously places the decision maker in a different situation. The difficulty of making choice is not due to a lack of information and is not eliminable in this sense, unless we interpret information so broadly that it covers realization by the decision maker of the relative importance he attaches to the various evaluations.

It can thus be seen that imposing rational conditions upon K for the case of multidimensional valuation has a lesser chance of success. This may explain that fact that no systematic endeavours have been made to do so, [10] even though there have been some proposals to indicate such criteria directly. They are usually criteria taken over from decision making under uncertainty, such as Savage's minimax of loss.

We shall not try to fill that gap. But it can easily be noted that if the utility functions u_i are to satisfy (4), then K would have to be limited by the following requirement:

(9) The set of optimal actions is invariant relative to the positive linear transformation of every function u_i taken separately.

This is an extremely strong requirement. Its analogue for decision making under uncertainty requires invariance relative to the same transformations of all the functions u_i. But (9) even implies the Rubin postulate, which is rather debatable (and stating the invariance of the optimal set relative to a change in the origin of the system of reference for every function u_i taken separately). Little could probably be added to (9) without running into contradictions.

This also expresses the difference between the two interpretations of (5).

Thus the term *optimization* is currently used to name processes which are essentially different from one another. There is, of course, nothing wrong in this, provided that we recognize these differences and mark their existence by a qualifying modifier.

In the strict sense of the term optimization means finding the answer to a closed question of the form: which element of A maximizes the function u_i? In the broader sense of the word we also speak about optimization when that question is not formulated unambiguously, either because the set A is defined in too general terms, or because one has maximization of more than one function u in view. In the latter case that unambiguous problem is made more precise by the requirements imposed upon its solution, and solution consists in reducing the issue to optimization in the strict sense of the term.

In all those cases which are not optimization *sensu stricto* we have to do with a solution which is arbitrary to some extent, that is, simply with a decision. This must be borne in mind, because the term *optimization* used in this way loses its fully objective nature. This applies both to a choice of action in an open problem and in the two above-mentioned cases of the plurality of the functions u. Such problems are not infrequently encountered.

NOTES

[1] R. D. Luce and H. Raiffa, *Games and Decisions*, John Wiley and Sons, 1957, Chap. 2.

[2] Ibid.

[3] Cf. O. Lange, *Optimal Decisions*, Polish Scientific Publishers, Warsaw, 1971, pp. 7 ff.

[4] The condition (3), or (3′), is disregarded as it is evidently dependent upon the two preceding ones. Weakening of any of those two cancels the said condition automatically.

[5] We have here practically to do with the Aristotelian scheme of defining: *per genus* (not necessarily *proximum*) *Z et differentiam specificam F.*

[6] J. Giedymin defines an open question thus: "Generally, a question for which we are not in a position to give a complete list of (a finite number of) admissible answers nor to give a scheme or an effective method of constructing admissible answers, will be termed an open question or one the answers to which are not

effectively constructible." Cf. J. Giedymin, *Problemy, założenia, decyzje* (*Problems, Assumptions, Decisions*), Polskie Towarzystwo Ekonomiczne, 1964, pp. 15–16.

[7] Cf. K. Szaniawski, 'Criteria of Decision Making', in: J. Kozielecki, (ed.) *Problemy psychologii matematycznej* ('*Problems of Mathematical Psychology*'), Polish Scientific Publishers, Warsaw, 1971.

[8] It is given here with an inessential reformulation, in order to obtain utility, and not negative utility, which would have to be minimized.

[9] Cf. K. Szaniawski, 'The Value of Perfect Information', *Synthèse*, Vol. 17, 1967, and 'Pragmatyczna wartość informacji' ('The Pragmatic Value of Information'), in: *Problemy psychologii matematycznej* (cf. note 7).

[10] This is so if we disregard the adaptation for that purpose of the postulates which describe collective decision making. We can treat those 'aspects relative to which action is evaluated' as persons who make such an evaluation. We can then apply the well-known set of postulates which is due to Arrow, but in such a case new troubles arise, to which Arrow has pointed himself. We also disregard the issue of an incomplete use made of the assumptions (preferences instead of utilities) and the problems of the adequacy of Arrow's postulates for decision making under the plurality of evaluations.

Eugeniusz Olszewski

ON NEW PRINCIPLES OF THE CLASSIFICATION OF SCIENCES*

Translated by Andrzej Ehrlich

I

The method of research has up to now been considered to be the main distinguishing factor of different scientific disciplines. Thus it is the method rather than the subject of research that has so far distinguished history from archeology or sociology, biochemistry from physiology, and astronomy from some branches of physics.

Recent tendencies to integrate research, however — characteristic of the period of the scientific-technological revolution — have diminished the usefulness of such a basis for classifying sciences. This is connected mainly with the so-called horizontal integration, which "consists in the interpenetration and overlapping of traditional disciplines, and the concentration of various kinds of research around complex problems". [1] Such complex problems are as a rule connected with a specific subject of research, which is so complicated and multi-aspective that its study requires the use of different research methods, i.e. — in the currently used classification — references to various scientific disciplines.

We might, therefore, conclude that the subject of research, and not the research method, would be a more suitable basis for classifying sciences in view of contemporary trends of development

* This article is an expanded version of the paper first published as 'O nowe zasady klasyfikacji nauk' in an internal bulletin of Warsaw Technical University *Prace Instytutu Nauk Ekonomiczno-Spolecznych Politechniki Warszawskiej*, 1974, No. 7.

of science. The science of science, which deals with scientific activity and its results, and which employs various research methods (e.g., sociological, psychological, economic, historical, etc.) is distinguished according to precisely that classification principle. Genetics — in the sense of a science dealing with heredity and change in vegetable and animal organisms — became such a discipline when cytological, molecular, and biochemical methods were added to statistical ones used originally.

If, however, we base the classification of scientific disciplines on the subject of research, it would no longer be autonomous, and would depend on the classification of reality as reflected by science. The principle of such a dependence was formulated a century ago by F. Engels who, as a basis for the classification of sciences, accepted the differentiation of forms of motion which exist in reality and are explored by science. [2]

II

Reality, which is reflected by science with an ever-increasing precision, is so complicated that its unequivocal and rigorous division would not be of much use for its cognition. All the following distinctions and systematizations, therefore, should not be considered to be classifications which adhere to the principles of formal logic, but only should be taken as typologies which are neither disjunctive nor complete.

In considering reality, it is possible either to distinguish its individual branches and fragments, or to observe aspects and characteristics of reality as a whole or of its major areas. The first step in systematizing scientific disciplines is, therefore, to divide them into the ones which explore specific aspects of reality as a whole or of its major areas, and the ones which explore given fragments of reality, i.e., given complexes of phenomena. [3]

The first group of scientific disciplines, which could be named aspective ones, includes in the first place materialistic dialectics; following that, logic (in the sense of Prof. T. Kotarbinski's definition, in which it is considered to be a science which, "attempting to ensure the correctness of our conclusions, examines the most general, timeless, and non-casual relationships between things, i.e., those relationships which should be reflected in our reasoning" [4]); mathematics, which explores quantitative and spatial relationships in reality as a whole; cybernetics, which explores control and feedback relationships, etc. [5]

Since science is a social activity closely linked with all human actions, the division of the second group of scientific disciplines, which we here call fragmentary disciplines, should be carried out from an anthropocentric point of view. Primarily, reality can be divided into the one formed without man's participation, and one formed by man — of course from matter supplied by nature — or at least formed with man's participation, i.e., carrying a mark of goal-oriented and planned transformation of matter through human activity [6]. Accordingly, fragmentary sciences are divided into natural sciences [7], and ones which might be called prospective sciences, because the subjects of their research belong — in cybernetic terminology — to prospective isolated systems.

Prospective sciences explore material and spiritual [8] creations which are results of human activity, and human activity itself [9]. Further systematization of these sciences should correspond to the systematization of human activities.

Human activities can be divided into informational ones — i.e. ones which take place in the sphere of information — and practical ones, which go on outside that sphere. Among practical activities one can distinguish economic and social ones. The first can be considered to be conscious and goal-oriented, because they are aimed at achieving some definite purposes by certain means. In this context, the means are things or material processes, and the goals are connected with those things or processes. [10]

The things and material processes may belong to the world of nature — which includes, among other things, human and animal bodies, and processes which take place in them — or in the world formed by man, or at least formed with his participation. Starting from this distinction, it is possible typologically to systematize economic activities: if things and processes connected with activities, as well as the means (or their majority) are man-formed, we are dealing with technological activities; in medical activities the means are usually also formed by man, but the goals are connected with natural things and processes, i.e., organisms and biological processes of man and other animals; in agricultural activities a great part of things and processes connected with the goals of activities, as well as the means, are natural things and processes, mostly of an organic nature. [11]

The human world comes into existence as a consequence of man's activity. It is composed of technological artefacts and processes (including means for furthering economic activities), of agricultural products, of desired changes in bodies and in biological processes of man or of other animals and, finally, of society. The latter is formed in the course of social activities, the goals and course of which depend to a large extent — as historical materialism has stated — on processes and results of economic and, in the first place, of technological activities. Information is created, transformed, and transmitted in the course of informational activity, and this activity, can be considered to be an integral part of society.

The division of prospective sciences should be a reflection of such a typological systematization of the human world. Among them, we might distinguish aspective-prospective sciences, connected with human activity as a whole. For example, axiology studies problems of determining objectives of activities, praxiology explores them from the point of view of effectiveness, i.e., the relationship between the objectives and the results of an activity, and of efficiency, i.e., the relationship between its costs and benefits. [12] Political economy explores the social principles of economic activity. [13]

Individual groups of economic activities and their results are examined by engineering, medical, and agricultural sciences; and social as well as informational activities and their results — by social sciences [14].

III

From the fact that prospective sciences (including aspective-prospective ones) are concerned with human actions and with the world formed by man — and, as the species *homo sapiens* develops, formed to an increasing degree in a conscious and goal-oriented, manner — we can draw conclusions concerning the relationship of these sciences to natural ones. We should take it that not only social sciences, but also other prospective ones examine higher forms of motion of matter than natural ones, as all prospective sciences are concerned with objects or processes which first appeared in human consciousness or which bear its mark. This leads to a separateness of the laws determined by these sciences, as "... the laws of lower forms of motion (though they must be taken into account) are not sufficient for explaining the higher forms".[15] Just as the laws of physiology cannot suffice to explain psychological phenomena, the laws of physics and chemistry are not sufficient for the explanation of phenomena which take place in objects and processes of technology.

The description of prospective sciences leads also to the pointing out of another significant difference between them and aspective and natural ones, namely the broader subject area covered by the former.

The basic task of science — as we mentioned earlier — is the reflection, or cognition, of objective reality. Every science answers more or less general questions: What is reality like? What is the world like? But the examination of man-dependent reality moreover poses other questions, questions which natural sciences, or mathematics and other aspective sciences, are free of. These are: Does human reality satisfy the goals for which it was formed? Does it properly satisfy the needs it was supposed to satisfy? These diagnostic questions lead to other ones: How should goals of activities be established so as to satisfy specific human needs in an optimal way? Or, in other words, what should be the changes of reality caused by man? What conditions should they conform to?

Accordingly, prospective sciences include not only cognitive research, but also evaluative research.[16] This results in the establishment of obligatory patterns for the effects of individual kinds of human activity, and for the activities themselves.

The most essential part of evaluative research is the determination of criteria for optimization, and the choice of criteria parameters which assume extreme values in optimal solutions. These criteria are derived from various fields of human activity. In the field of technology, their nature may be technological in character (e.g., minimizing waste in a refrigerator), economic (e.g., minimizing costs) or medical (e.g., minimizing environmental pollution). In practice, it is necessary simultaneously to take into consideration criteria derived from different fields of activity.

Since goals attached to human activities and to their effects depend on the system of values which exists in a given society, the choice of optimization criteria must depend on the same factors. Now, since the system of values in a class society to a large extent depends — as is well known — on the social structure existing in the given country and on the governing class, we may conclude that the class character of prospective sciences, and in particular of social ones, is much more marked than that of aspective or natural ones.

This relationship in all scientific disciplines influences the choice of research programmes, methods, and hypotheses. However, prospective sciences include evaluative research as an additional element which is always related to value systems, and therefore evaluation problems always have to a lesser or greater degree a class character. This is obvious in the case of social sciences, but it is also true, either in a specific or in an indirect way, of other prospective sciences. For example, optimization criteria for technological conditions from the point of view of work-safety or of the state of health of the population may be formulated differently in capitalist and socialist countries; significant examples can easily be found in countries which apply a policy of apartheid, where such optimization criteria are different for the 'white' and 'coloured' population.

IV

The general typological proposals presented in this paper might, it seems, serve as a useful basis for the development of a classification of sciences which would be well suited to the contemporary stage of development of science linked to the development of technology in the process of the scientific-technological revolution.

This revolution brings, on the one hand, an increasing number of scientific disciplines to such a high level that they become applicable, and can therefore not only study and interpret reality, but also contribute to its changes; and on the other hand it develops the technological capacities of man to an extent which enables him to go on, from using nature for his ends and taming it, to creating a new world of his own, composed of the artefacts and processes of technology. [17] This required the formation and development of sciences, called prospective ones in this paper, which, as mentioned previously, not only answer questions of the type: What is the world like? but also of the type: To what requirements should the world formed by man correspond? — which therefore formulate scientific premises and conditions for the formation of the human world.

Now prospective sciences — with the exception of some social ones — began to appear only in the beginning of the 19th century, following which they developed at an ever-increasing rate, parallel to the increasingly magnificent achievements of technology, medicine, and argiculture. Born late, they did not gain a suitable place in the classification of sciences, particularly so as they were often treated as so-called 'applied sciences', as it were, a dependent addition to natural sciences, deprived, moreover, of cognitive tasks. [18]

A classification based on the presented principles could ensure a suitable place for prospective sciences in the whole system of sciences, pointing out, moreover, the similarity of the scope of research and its relationship to the value system of all prospective sciences, social as well as technological, medical, and agricultural. This leads, among other things to the conclusion that the concepts presented may have not only a theoretical significance, but may also serve as a useful tool in the formation of scientific policy, and in planning research and evaluating its results.

NOTES

[1] See: I. Malecki, E. Olszewski, 'Regularities in the Development of Contemporary Science'. In: B. Barnes (ed.), *Sociology of Science*, Harmondsworth, England, 1972, p. 153.

[2] See: B. M. Kedrov, 'The History of the Classification of Sciences', *Organon*, 1964, No. 1, pp. 176–178. B. M. Kedrov applies the Engels principle in his numerous papers on theory of science; cf. his three papers in the Russian monthly *Problems of Philosophy* (1972), No. 2 and No. 10; (1973), No. 3.

[3] See: E. Olszewski, 'Rozważania o pojęciach nauki i techniki' ('On notions of science and technology'), *Archiwum Inżynierii Lądowej* (1972), Nos. 3–4, p. 450.

The idea of the basic division presented here was caused by Dr. A. Gella's remark at the international symposium on general problems of history of science and technology, organized in 1963 in Jabłonna (near Warsaw) by the History of Science and Technology Institute of the Polish Academy of Sciences. Cf. *Organon* (1964), No. 1, p. 194.

[4]. See: T. Kotarbiński, *Leçons sur l'histoire de la logique*, Paris, 1964, p. 208.

[5] B. M. Kedrov includes mathematics and cybernetics — but not dialectics and logic — in 'abstract sciences'; cf. the last of the papers quoted in the note 2.

[6] See: E. Olszewski, 'O pojęciach techniki i nauk technicznych' ('On the concepts of technology and engineering sciences'), *Zagadnienia Naukoznawstwa*, No. 3 (1970), p. 10.

[7] The further break-down of natural sciences will not be discussed here.

[8] The term 'material' is used here as the opposite of the term 'spiritual' understood as 'connected with human consciousness'.

[9] It is worth mentioning that physiological acts should not be treated as a part of human activity, as they depend on the human organism — which is not a result of man's planning but a creation of nature — and on processes which take place outside of human consciousness to such an extent that they should rather be included in the world of nature, which is independent of man.

[10] See: O. Lange, *Political Economy*, Vol. 1: *General Problems*, Oxford–Warsaw, 1963, p. 132–133.

[11] See: E. Olszewski, 'Rozważania...' (*op. cit.*), p. 446.

[12] Organization and management science can be considered to be a part of praxiology defined in this way.

[13] See: O. Lange, *op. cit.*, p. 16.

[14] It might be useful to exclude 'informational sciences' (i.e., science of science, science of designing, and didactics — in the broad sense of the term) from the group of social sciences.

[15] W. Krajewski, 'Mechanicyzm i redukcjonizm' ('Mechanism and reductionism'), In: *Z dziejów mechanicyzmu w fizyce i chemii* (*On the History of Mechanism in Physics and Chemistry*), Wrocław, 1974, p. 16.

[16] In contrast, the author considers as unjustified the inclusion of problems of realization, i.e., of establishing different types of rules for practical activity in the scope of scientific research. In his opinion this is the task of practical (e.g., technological) research, which cannot be identified with scientific research; cf. E. Olszewski, 'O pojęciach...' (*op. cit.*,) pp. 9–10.

[17] See: E. Olszewski, 'Les sciences et les techniques dans la période de la révolution scientifico-technique', *Organon*, No. 8 (1971), p. 46–49.

[18] See: E. Olszewski, 'The Substance of Engineering Sciences and their Role in the Period of Scientific and Technological Revolution', In: *Methodological and Social Problems of Technology and Engineering Sciences* (in Russian), Leningrad, 1972, pp. 34–35.

Marian Mazur

THE MORPHOLOGICAL STRUCTURE OF SCIENCE*

Translated by Christopher Kasparek

Introduction

The social importance of science has given rise to a tendency to develop science in an organized manner, the chief manifestation of this being the financing of scientific research in order to secure the greatest possible benefits from the results obtained. Little wonder that this is accompanied by the development of studies of science itself.

A widespread but thoroughly mistaken approach to creating knowledge about science is to begin by considering the meanings of words, a process that seeks answers to such questions as, 'What is science?','What is the science of science?', etc. The answers are dictated by the respondent's personal linguistic habits, and this leads to citing definitions given by various authors, to imputing 'erroneousness' to some and to championing or proposing others. This in turn results in barren disputes over whether something is or is not 'scientific', over whether something does or does not belong in 'the science of science', and so on, as though the subject matter were to depend on the terminology, rather than vice versa. By the same token, many important matters have been overlooked simply because of a lack of terms to designate them.

The present paper takes as its point of departure certain elements

* Written in 1977, previously unpublished.

of realia for which terms are stipulated. Clearly, such terms cannot be 'wrong' but at most inconvenient, and for that reason may be questioned, and demands may be made for their replacement with other terms, but this in no way alters the realiá that are here considered.

And the realia are such that — just as there is an activity that begins with the arrival of workers to an empty site and ends in their departure forever after they have completed objects that assure suitable conditions for human habitation, and this state of things is not altered whether the objects be called 'houses', 'buildings', 'structures' or something else — so there is an activity that begins with questions that have not been answered by anyone (ignorance) and that ends with exhaustive answers (knowledge); and such activity does not depend on the terminology that is applied to it.

It is stipulated that the creation of knowledge will here be designated as 'science' (investigative activity), while the specialists who occupy themselves with creating that knowledge will be designated as 'scientists' (investigators).

In addition to science as the creation of knowledge, we may distinguish the creation of knowledge about science itself — this we shall designate as 'the science of science', or 'logology'.

In order to proceed from ignorance to knowledge, one must carry out several tasks (scientific tasks, investigative tasks). Again, in order to carry out these tasks, one must make use of suitable resources (scientific resources, investigative resources). The successful creation of knowledge (development of science) depends on congruity between the tasks carried out and the resources applied.

Investigative tasks

If one arranges the tasks that occur in the process of creating knowledge in logical sequence, that is, in such an order that only after a previous task has been executed can one go on to the next, then one will obtain the discrete stages in that process. Only in an ideal case, i.e., given unfailing aptness of procedure at the various stages, would this also constitute the chronological sequence. In creative activity, however, this happens seldom for, due to lack of precedents not everything can be anticipated. In the later stages something may turn up that will require a return to earlier stages, although this does not change their logical sequence.

(1) *Problem.* Before there can be an answer to a question, the question must first be formulated. Hence the first stage is to pose the problem.

Usually posing a problem is regarded as being the same thing as stating its title, with at the most a brief explanation being given as to what is involved in the problem. In the belief that this is a matter not requiring any investment of effort worth mentioning, it is not regarded as being a stage at all. Yet it involves serious preparatory work encompassing familiarity with the current status of knowledge, familiarity with existing needs, the setting of priorities, and the choice of a problem that is feasible of solution. Neglect in this respect often leads to the selection of problems that have already been solved, that are of little consequence, that are hazy, that are impossible to solve given the available resources, and so on.

(2) *Solution.* After the problem has been posed, the next step is to solve it; this is the second stage in the investigative process.

The simplest procedure is to apply the most appropriate of known methods. Only when none of these prove useful, does the scientist or scholar face the necessity of elaborating a new method.

The choice of method is made difficult by the fact that the various methods are scattered about in diverse publications. An exception are textbooks, but these quickly become outdated and generally at the time of their writing contain only methods that are already regarded as being classical. Indeed, a useful purpose would be served if scientific methods were catalogued and published in frequently updated lists.

(3) *Presentation.* The third stage is presentation, i.e., making the solution available to those to whom it might be of interest.

This third stage, like the first, is often so far unappreciated as not to be regarded as being a stage at all, due to the mistaken notion that the author ought to describe the solution in the very course of obtaining it and thus that the conclusion of the second stage ought to yield not only the solution but the text as well.

Yet the third stage is indispensable and nearly always time-consuming. Quite apart from what the author writes in the process of solving the problem, he will still have a lot to do later with respect to style, standardization of terminology and notation, review of the literature, purpose, conclusions and recommendations. Still more important, but as a rule disregarded, is a solicitude for the needs of the reader, a concern for writing the text in a manner that will make it comprehensible to him. This is an exceedingly important matter, inasmuch as conventional text by itself is now hardly ever adequate in any scientific discipline. More and more, use must be made of such means of presentation as logical symbols, mathematical formulas, diagrams, tables, systems flowcharts, graphs, etc. Here again, it would be useful to study and compare the types of matter to which the various means may profitably be applied, in order to make scientists better aware of which means are replaceable by others and to guide school programmes in preparing future readers of scientific literature.

Furthermore, a scientific or scholarly text should make clear just what are the innovations it offers. After all, it is only for the sake of the anticipated innovations that the prospective reader reaches for the text of a scientific or scholarly investigation.

And yet authors, instead of stating clearly what innovations their research introduces, proceed as if desirous of effacing the extent of their own contributions. This approach stems from traditional science, which was dominated by a keen sense of the contrast between the noble disinterestedness of scientists (the statement, 'I discovered this', would have inspired distaste as an expression of immodesty) and the pecuniary cupidity of inventors (the statement, 'I invented this', has been commonplace in patent litigations).

(4) *Innovation.* Regardless of the author's attitude toward defining his own innovations, these must be determined by other scientists or scholars who have been called upon to review his work. This is the fourth stage in the investigative process.

If the author has stated the nature of his innovations, the

reviewer's task is thereby made easier, for he then has only to confirm or to question the author's assertion.

In the absence of a declaration by the author concerning his innovations, the reviewer is in a difficult situation. In order to state whether the work contains something new, and just what that might be, he would need to study the report thoroughly, which usually he has neither time nor inclination to do. A superficial reading courts ascribing nonexistent innovations to the author or denying actual innovations. As a result, reviewers generally see their task in terms of rendering such judgments as that the paper is valuable, extensive, bespeaks the author's broad knowledge, etc. — judgments that are often prompted by a desire to promote the author to the dignity of doctorhood, rather than society's interest in innovations.

(5) *Application.* The fifth and final stage in a scientific or scholarly investigation is to demonstrate the possible applications of the innovation (which was presented in the third stage and confirmed in the fourth).

This stage is sometimes an object of dispute concerning who is responsible for implementing innovations — scientists or users? The disputes arise from subsuming under 'implementation' both the route from innovation to application (the province of scientists) and that from application to exploitation (the domain of the users of scientific research), without a strict distinction being drawn between the two.

Application and exploitation need not involve technological innovations. They may equally well pertain to innovations in the form of information that can serve exclusively the satisfaction of someone's curiosity (e.g., to an archeological discovery). In such cases the application is popularization — the transformation of a scientific or scholarly text into a text that is suitable for the general reader — with the popularization properly being carried out by scientists or scholars as the fifth stage of investigative work. The exploitation, finally, is education — the teaching of what has been popularized: an assignment not for the scientists or scholar but for the user (lecturer, teacher).

Resources for research

In order to carry out his tasks, the scientist or scholar must have access to the requisite resources. The latter can be placed in an order, beginning with those that are most closely bound up with the scientist — his personal traits — and going on to ones that are increasingly loosely associated with him: the equipment he uses, and the environment in which he is obliged to work and on whose help he can count. From this standpoint, one may distinguish nine kinds of resources for research.

(1) *Talents*. Most intimately bound up with a scientist are his innate qualities — his talents. These are resources that the scientist possesses even before he becomes a scientist.

At first blush it may seem that talents are not a resource amenable to organization. This view, however, is true only with respect to individual persons: if the latter lack the required talents, then obviously no one can supply them with those talents. It is not, however, true with respect to science as a profession, for it is possible to select candidates for that profession likewise from the standpoint of their talents.

Nevertheless, up to now science has not been a profession to which candidates are selected for their talents, because up to now no one has succeeded in defining the talents that should be required (except in quite general terms such as inquisitiveness, a critical mind, inventiveness, etc.), much less how they might be tested for.

(2) *Skills*. In addition to talents as innate qualities, acquired investigative skills affect an individual's aptitude for the scientific profession.

In traditional science and scholarship, skills of this kind used to be acquired through instruction by experienced scientists or scholars and through observation of their work, as well as through one's own investigative practice. Nevertheless, this process was largely accidental, fragmentary and long-drawn-out.

In order to avoid these shortcomings, attempts are made to organize the education of candidates for the scientific profession but no one yet knows what that education should consist of. Doctoral studies have been intended as one means of acquiring investigative skills, but it is easier to appropriate funds for doctoral scholarships than to define what their recipients should do. From the scientific standpoint, the doctoral student ought to work creatively, preparing a doctoral dissertation, but the administrative point of view is different: since the doctoral candidate is paid a stipend, he should be watched so that he will not merely take the money and do nothing. And from there it is a straight line to saddling him with lectures, checking attendance, and holding trivial examinations. As a result, the doctoral student becomes a sort of glorified undergraduate.

(3) *Form.* The next personal resource of the scientist or scholar is his form — his unceasing physiological and psychic readiness for scientific work, encompassing health, peace of mind, and freedom from mental distractions.

It is paradoxical that up to now the expression 'form' has generally been applied only to athletes. Solicitude for the form of the latter goes very far: special funds to keep them in tip-top form, special medical care, special training camps prior to important competitions, various benefits and privileges, and so on, just so that they will win valuable points and medals manufactured out of variously coloured metals.

But no one concerns himself with scientists's form, even though their accomplishments are certainly more useful than are the athletes' records. The results of this unconcern principally affect the form of young scientists and scholars, those at the biological peak of their creative capabilities (between 30 and 40 years of age).

There are no grounds for extending unsecured credit to beginning scientists and scholars, but once it is determined that they are in fact genuine scientists or scholars, if young ones, they should be assured a radical improvement in their living conditions.

(4) *Information.* After the personal resources of the scientist or scholar (talents, skills, form) we should next consider those resources that constitute his external equipment. First and foremost is a flow

of information enabling him to learn of the newest achievements of learning in the world, and to make use of these.

(5) *Apparatus*. In addition to information, the scientist or scholar needs technical, for example measuring, apparatus. Until recently, this kind of resource was basically involved only in disciplines that require laboratory experimentation. In the present state of computer technology, few disciplines can do without scientific apparatus.

(6) *Helpers*. The scientist's resources may also be said to include his helpers. The latter ought to carry out tasks on which highly qualified scientific workers should not have to waste their time and effort: routine measurements, calculations, diagrams, records, proof reading, indexing, mimeographing, xeroxing, taking dictation, expediting simple correspondence, etc.

(7) *Institution*. In moving on from the scientist's equipment (information, apparatus, helpers) to environments considered as resources for scientific or scholarly work, we must mention first of all the institute (meaning here any separate research institution), which constitutes the environment most closely associated with the investigator.

The institute as resource for scientific work is an organization that facilitates scientists' investigative tasks.

(8) *Cooperation*. An environment associated with the scientist or scholar more loosely than the institute comprises all remaining scientists or scholars. Cooperation with these is likewise a resource for the scientist's or scholar's research efforts.

A typical ground for cooperation among scientists and scholars used to be scientific societies, while at present an increasing role is played by scientific and scholarly conferences. And these are undergoing characteristic changes, clearly associated with the growing numbers of scientists and scholars and of their publications. The number of conferences, domestic as well as international, is growing by the year, so that the interested scholars do not learn about all of them in time. The length of presentations, the time for their delivery, and time allotted for discussion are increasingly being limited. These limitations notwithstanding, scientific and scholarly conferences are a very valuable medium of cooperation: they assure dissemination of the newest ideas in a given discipline and facilitate personal contacts, private discussion of details, exchange of publications, etc.

(9) *Society.* The environment most loosely associated with the scientist is society in the broadest sense of the word (it can even encompass all humanity). Society figures here as a resource for scientific work in view of the latter's clear dependence on social support.

This support likewise changes even as science develops, changes often for the worse. This is evident in accusations that scientists have failed to solve the great social problems, that their achievements have increased the cruelty of war, and that they are contributing to the degradation of man's natural environment through air, water, pesticide, noise and other pollutions. A growing chorus of voices argue that science is becoming increasingly expensive: and the argument is being conducted in a tone implying that government expenditures for science are some kind of charity or a form of self-denial borne for the sake of science.

These are obvious misapprehensions. Scientists create new ideas and possibilities, but it is not they who decide about the applications and exploitation. As far as the expenditures on science are concerned, these are borne primarily because science happens to be one of the most profitable enterprises ever to have been devised by mankind.

The matter is aggravated by the fact that the critics of science are not in the least disposed to listen to what scientists might tell them. Sports events draw millions to the television screens; how many people attend scientific or scholarly lectures?

The isolation of scientists from society is also deepening because science is becoming increasingly difficult to understand for persons lacking in adequate preparation, and it is hard to be enthusiastic about something that one does not understand.

The morphology of science

The foregoing discussion has highlighted 5 kinds of investigative tasks,

1. the posing of problems,
2. the solution of problems,
3. the publication of solutions,
4. the determination of innovations,
5. the application of innovations;

and 9 kinds of resources for research:

1. talents,
2. skills,
3. form,
4. information,
5. apparatus,
6. helpers,
7. institution,
8. cooperation,
9. society.

Together these constitute 14 factors that must be considered if the effectiveness of investigative activity is to be maximized.

One must also remember the necessity of differentiating among the resources with respect to the various tasks. For example, different talents are needed for the posing of problems than for their solution, and still others are needed to apply innovations, and so on.

The detailed differentiations can be obtained by associating all the resources with all the tasks. In this way we obtain 45 components of science treated morphologically, as in Table 1.

The virtues of a morphological formulation of the structure of science include its completeness, which assures that none of the components of the structure will escape our field of vision.

Another of the system's virtues is its disjunctiveness, which assures that if something belongs in one of the components, then it does not belong in any other.

TABLE 1

	Problems	Solutions	Presentations	Innovations	Applications
Talents					
Skills					
Form					
Information					
Apparatus					
Helpers					
Institution					
Cooperation					
Society					

A morphological treatment of science may also contribute to greater order in logological (science-of-science) literature by directing it more toward the needs of science rather than toward the authors' own interests. For example, when publications appear on 'the psychology of science' or 'the sociology of science', it is evident that the authors have written them because they are psychologists or sociologists and wish to share with readers their own knowledge as applied to science. That does not, however, make clear just what needs this approach is supposed to satisfy. To give an example, the solving of problems by a groups of scientists at an institute is something quite different from cooperation among scientists in determining the innovations in someone's work, even though psychological and sociological aspects do enter into the picture in both cases.

Setting priorities for science-of-science problems

Until now, the selection of logological (science-of-science) problems has been essentially accidental in character. However, by using the morphology of science, it is possible to introduce elements of value judgment into it, e.g., by positing that the priority for studying a given problem shall be the higher,

— the greater the importance of the problem, or

— the less the problem has been studied (in other words, the greater the dearth of papers on the subject).

In order to make possible a numerical designation of priorities, let us adopt, both for the importance of a problem area and for its current degree of neglect, a 5-point scale of ratings, 0–1–2–3–4, to be used in calculating the products of ratings, these products being treated as resultants of the priority ratings of the various problem areas.

For practical purposes we may reject a '0' rating of importance (a completely unimportant problem area) as well as a '0' rating of neglect (a completely exhausted problem area), considering such cases to be improbable. For similar reasons we may also reject the opposite extremes, i.e., a '4' importance rating (an absolutely important problem area) and a '4' neglect rating (an untouched problem area).

Upon such a reduction, we obtain a 3-point scale of ratings for importance — 1 (small), 2 (medium), 3 (large) — and for neglect: 1 (small — there are many papers), 2 (medium), 3 (large — there are few papers).

All the possible products of such ratings form the following scale of resultant priority ratings: 1–2–3–4–6–9. An optional further operation, useful from a practical standpoint, will be to arbitrarily augment the '9' ratings by 1 point, since the double-digit number 10 is visually more prominent and thus makes problems of highest priority immediately noticeable. This may be done without prejudice to interpretation, since the scale of priority ratings is already discontinuous anyway in its upper portion (lacking as it does a 5, 7 and 8).

Sample ratings for the importance of various logological problem areas are given in Table 2.

TABLE 2

Importance ratings

	Problems	Solutions	Presentations	Innovations	Applications
Talents	3	3	2	2	3
Skills	3	3	2	2	3
Form	3	3	1	1	3
Information	3	3	1	3	2
Apparatus	2	3	1	1	2
Helpers	1	3	1	1	3
Institution	2	3	1	1	3
Cooperation	1	2	1	3	2
Society	2	1	1	1	3

Sample neglect ratings for the various logological problem areas are given in Table 3.

Multiplying the importance ratings by the neglect ratings, as taken from the foregoing tables, we obtain products that constitute

TABLE 3

Neglect ratings

	Problems	Solutions	Presentations	Innovations	Applications
Talents	3	3	2	2	3
Skills	3	1	2	3	3
Form	3	3	3	3	3
Information	1	1	1	1	2
Apparatus	2	1	1	3	2
Helpers	3	1	1	3	1
Institution	3	1	1	2	1
Cooperation	2	2	1	1	2
Society	3	1	2	1	2

priority ratings for the various logological problem areas. These are given in Table 4.

TABLE 4

	Problems	Solutions	Presentations	Innovations	Applications
Talents	10	10	4	4	10
Skills	10	3	4	6	10
Form	10	10	3	3	10
Information	3	3	1	3	4
Apparatus	4	3	1	3	4
Helpers	3	3	1	3	3
Institution	6	3	1	2	3
Cooperation	2	4	1	3	4
Society	6	1	2	1	6

The priority ratings thus obtained result from a formal operation (multiplication), but the data used in that operation — the importance and neglect ratings — are of course encumbered with the arbitrariness of the persons who performed the ratings.

In the present paper the ratings have been carried out by the author, who trusts that his ratings will be corrected by other workers, who may perhaps hold broad discussions on the basis for the ratings.

In any case, the arbitrariness of individual ratings is strongly tempered by the fact that each of the problem areas is subject to two criteria at once: tasks and resources. Thus the importance rating for each problem area must bear comparison both with the importance ratings of other problem areas for the same kind of task and with the importance ratings of problem areas for the same kind of resource. In other words, the importance rating for a given problem area appears at the intersection of a certain row and a certain column. Consequently the role of an importance rating among importance ratings for various tasks cannot be at odds with the same rating's role among importance ratings for various resources. The inadmissibility of such incompatibility thus endows the described procedure with a largely self-correcting character.

For example, a high (3) importance rating for cooperation among scientists in determining innovation means that such cooperation is considered to be important as compared with other resources for determining innovation. At the same time, this rating means that such cooperation is regarded as being important in comparison with cooperation in other tasks. Only in the event that both these views are simultaneously considered to be justified, can this rating be upheld.

It is a similar case with neglect ratings, except for the added circumstance that the state of the literature is objectively verifiable. For example, one can hardly maintain that the neglect of a certain problem area is small (that there are many papers in it) if he can point to no papers in that area.

The procedure discussed is one of value analysis, inasmuch as priority ratings are arrived at not by comparing them for various problem areas but exclusively on the basis of previous importance and neglect ratings. A problem area whose place on the priority list seemed at odds with intuition could not be moved to a lower or higher place otherwise than by revising the importance and neglect ratings for the problem area. If such a review led to keeping the two ratings, then the priority rating too should be maintained, and the intuition regarded as having in that particular case been mistaken.

Going back now to the derived table of priority ratings, it is worth noting that the quantitative distribution of the 45 ratings contained in it is as follows:

Priority rating	No. of problem areas
10	8
6	4
4	8
3	15
2	3
1	7

What is immediately apparent here is that the greatest number of problem areas have a 3 rating. Such a rating can only result from multiplying a 1 by a 3 or a 3 by a 1. That means that a 3 rating takes in problem areas of little importance and great neglect (few papers) and problem areas of great importance and little neglect (many papers). But that is a normal situation: if something is of little importance, then little is written about it, and if something is very important, then a great deal is written about it. Therefore, if the science of science were developing perfectly normally, all its problem areas would merit a 3 rating. Consequently, what is amazing is not that the greatest number of problem areas do have this rating, but that their number is still so small. The difference between the number of such problem areas and the total number of problem areas is, as it were, a measure of the abnormal state of the science of science.

In the setting of priorities, the most interesting are of course the problem areas with the highest priority rating. Such a rating can result only from multiplying a 3 by a 3, which means that these are problem areas that are important but neglected, that is, problem areas that have been overlooked. The table of priority ratings shows that the problem areas in question involve the role of scientists' personal qualities (talents, skills, form) in the posing and solution of problems, and in the application of the innovations created. In the preoccupation with equipment and institutions, the influence of scientists' personalities has been overlooked.

This is less surprising than might seem, if one considers the fairly widespread conviction among science administrators that the effectiveness of scientific research depends more on organizational rigours than on an awareness of the identities of individual scientists.

The factography of science

In addition to kinds of tasks ('the theory of science') and kinds of resources ('the technology of science'), increasingly there is a felt need to treat tasks and resources as facts ('the factography of science') occurring in a stated time ('the chronography of science') and in a stated place ('the topography of science'). A special role is played by the counting of these facts ('the factometry of science', also clumsily called 'scientometry', a term that creates the mistaken impression that what is involved is the measurement of scientific ideas), with respect both to time ('the chronometry of science') and to place ('the topometry of science').

Of some importance is the correlation of the facts of science or scholarship with time; in connection with this, one may distinguish the following departments in the chronography of science:

(1) The history of science (the reconstruction of bygone facts).

(2) The diagnostics of science (the determination of current facts).

(3) The prognostication of science (the prediction of future facts).

(4) The programming of science (the recommendation of future facts).

Much more rarely felt is a need to correlate the facts of science or scholarship with places (e.g., the locations of scientific or scholarly institutions at home and abroad, the concentration of scientists or scholars in particular regions, etc.).

The practical importance of a factographic treatment of science rests in the fact that it is easier to make valid decisions concerning the tasks and resources of science when one knows what they have been, are, probably will be, and should be like.

The morphological classification of science-of-science literature

The morphology of science may also serve as a foundation for the classification of logological literature. If all logological publications were written in a manner that clearly indicated what components of science they were concerned with, then one could use for that purpose the same table as has earlier been discussed, with the rows and columns additionally being numbered. There are, however, many publications that refer to various components of science or that treat them in a general manner. For such publications we must add another row (for publications that do not distinguish kinds of investigative resources) and another column (for publications that do not distinguish kinds of investigative tasks). The additional row and column may be designated by the digit 0. At their intersection is needed a space designated 00 (for publications of such general content that they permit of distinguishing neither tasks nor resources).

A table thus modified is given below, filled out with sample data collected by the author for the Science of Science Committee of the Polish Academy of Sciences, chiefly on the basis of an analysis of the problem areas discussed in papers published in issues 21–40 of the quarterly *Zagadnienia naukoznawstwa* in the years 1970–74, and encompassing a total of 139 papers.

TABLE 5

	1	2	3	4	5	0	
	Problems	Solutions	Present-ations	Innova-tions	Applica-tions	General tasks	Total
1 Talents		3	3	3		1	10
2 Skills		6		1		3	10
3 Form			2			6	8
4 Information			8	1	1	6	16
5 Apparatus		3	1			1	5
6 Helpers							
7 Institution		2		3		16	21
8 Cooperation			4			4	8
9 Society	1			1	2	21	25
10 General resources	3	9	2	4	1	17	36
Total	4	23	20	13	4	75	139

The data contained in the above table lead to the following conclusions.

As far as the tasks of science are concerned, the numbers given seem to indicate that logologists ('scientists of science') are not very interested in where scientific problems come from (only 4 papers in column 1, 3 of them general) or where they lead (only 4 papers in column 5, one of them general and two in row 9, easily suspect of the superficiality that is so often found in discussions on the social role of science). The top place, instead, is held by methodological matters (23 papers in column 2) and matters involving publication (20 papers in column 3).

As far as the resources of science go, the striking thing is the complete lack of papers on the subject of auxiliary personnel (an empty row 6). Perhaps this is so because, in the absence of helpers, their duties are shuffled off onto teaching assistants, who are thus treated as technicians — often the most convenient expedient for the professors of various grades who supervise their labours; and since it is mainly the professors who author the logological papers, it is hardly surprising if the matter of helpers is not treated as a problem.

Among the impoverished problem areas are living and working conditions (8 papers in row 3, 6 of them general) and cóoperation among scientists and scholars (8 papers in row 8, 4 of them general).

On the other hand, most frequently discussed are the organizational problems of scientific institutions (21 papers in row 7) and the social problems of science (25 papers in row 9, but 21 of that number general).

In summary, it may be said that the science of science concentrates on methodology and publications (but scientists and scholars have always been interested in methods of solving problems and in publishing the solutions), and on scientific and scholarly institutions and the connections between learning and society (which is no doubt connected with the appropriation of considerable social funds for scientific and scholarly institutions — a matter of particular interest to administrators). Jointly that comprises the subject matter of 83 papers, or 60% of all those analyzed. This would seem to suggest that the science of science is not yet an integrated discipline but is developing only in certain directions, mainly under the influence of scientific and scholarly traditions and the needs of science administrators.

The classification of scientific and scholarly literature also requires its categorization. As is evident from the modified table for the morphology of science, this requires 60 classifications, numbered bi-digitally. If it be stipulated that the first digit shall refer to resources and the second to tasks, then for example a publication devoted to scientific apparatus ought to be classified in the category numbered 52. A general publication on the subject of cooperation among scientists or scholars would enter category number 80. A general publication on the subject of innovations would have category number 04, etc.

A morphology of science likewise makes possible a classification that will take account of the factography of science, for example chronography. For this purpose digital designations may be introduced: 1 — the history of science; 2 — the diagnostics of science; 3 — the prognostication of science; 4 — the programming of science; and 0 — papers with no chronographic restriction. Thus, supposing it is agreed that the digit referring to the chronography of science will be given first place, we obtain a transformation of the bi-digital into a tri-digital numeration. Then a publication discussing the current status of research apparatus would have classification number 252. A publication devoted to the history of scientific and scholarly conferences would take classification number 180. A general publication on innovations without chronological restriction would receive classification number 004. A general publication on the prospects of science would have classification number 300. A most general publication on science or scholarship would be classified 000.

NOTES ABOUT THE AUTHORS

GASPARSKI, Wojciech W., b. 1936, graduated from Silesian Technical University in 1959, became a Doctor of Technical Sciences in 1969, and since 1971 he has been Associate Professor and head of the Design Methodology Unit in the Department of Praxiology and the Science of Science in the Institute of Philosophy and Sociology of the Polish Academy of Sciences.

His main interests are: the praxiological approach to research, the systems methodology, and the methodology of design.

He is a member of numerous Polish and international scientific organizations, e.g., the Committee of the Science of Science (Scientific Secretary), the Design Research Society (UK), the Design Methods Group (USA), and of Editorial Boards, e.g. the *International Journal of General Systems* (USA/UK), *Design Studies* (UK), *Zagadnienia Naukoznawstwa* (Poland), *Prakseologia* (Poland), *Metamedicine* (Holland/Federal Republic of Germany).

He is the author or editor of several books (in Polish, Russian, and English) and of numerous papers presented to international conferences, e.g., 'Praxiological Analysis of Design' (in Polish), *Praxiological Studies* (in English, co-editor with Tadeusz Pszczołowski). He also has organized several scientific conferences.

KACZMAREK, Jan, b. 1920, graduated from Cracow Technical University in 1947, became a Doctor of Technical Sciences in 1953 and since 1956 has been Professor at Cracow Technical University as well as at the Institute of Fundamental Technical Problems as head of the Department of Dynamics and Physical Properties of Surface Layers. Simultaneously he is the Scientific Secretary of the Polish Academy of Sciences.

His main research interests are: metal machining, as well as general problems of the science of science and science policy.

In 1962 he was elected a Corresponding Member of the Polish Academy of Sciences and in 1971 a Full Member; in 1960 he became a Fellow of CIRP (Collège International pour l'Etude Scientifique et Technique de Production Méchanique); he is also a foreign member of the following Academies: the Bulgarian Academy of Sciences, the National Academy of Engineering (USA), and the Royal Belgian Academy of Arts, Science and Literature. He has been granted a doctorate *honoris causa* by Bauman University of Technology in Moscow and the University of Technology in Karl Marx Stadt (German Democratic Republic).

He is the author, or co-author, of fourteen books (among them monographs in Polish: *Podstawy doboru warunków skrawania metali* (*Fundamentals of Selecting the Parameters of Metal Machining*), 1953; *Podstawy skrawania metali* (*Fundamentals of Machining*), 1956, and of over a hundred papers and treaties, several of them in English, French, German or Russian.

KOTARBIŃSKI, Tadeusz (1886–1981), was one of the most interesting contemporary Polish philosophers. He was Professor at Warsaw University, a member of the Polish Academy of Sciences (its President from 1957 to 1962), a Foreign Member of the Academy of Science of the USSR, a member of the Institute International de Philosophie (its President from 1960 to 1963), a member of the Institute de Philosophie Politique (Paris).

Praxiology, i.e., the theory of efficient action ('good work'), a branch of learning of which he is largely the creator, had been the main field of his theoretical research in the last years.

His main books are as follows: *Utylitaryzm w etyce Milla i Spencera* (*Utilitarianism in the Ethics of Spencer and Mill*), 1915; *Elementy teorii poznania, logiki formalnej i metodologii nauk* (*Elements of the Theory of Cognizance, Formal Logic and Methodology of Sciences*), 1929, 1961; published by Pergamon and Ossolineum as *Gnosiology. The Scientific Approach to the Theory of Knowledge*, 1966; *Wykłady z dziejów logiki* (*Lectures on the History of Logic*), 1967; French lang. ed. *Leçons sur l'histoire de logique*, 1965; *Traktat o dobrej robocie* (*Treatise of Good Work*), 1955, translated into English and published by Pergamon Press as *Praxiology*: *An Introduction to the Sciences of Efficient Action*, 1965; *Wybór pism* (*Collected Works*), two volumes, 1957, 1958.

MALECKI, Ignacy, b. 1912, graduated from Warsaw Technical University in 1935, became a Doctor of Technical Sciences in 1945 and since 1950 has been Professor of this University. Since 1957 he has been a Full Member of the Polish Academy of Sciences and a member of its Praesidium. He is also Director of the Institute of Fundamental Technical Problems and President of the Committee of the Science of Science in Poland.

His researches cover electroacoustics and the general problems of science, the methodology of science, and science policy.

During the years 1969–1972 he was Director of the Science Policy Department of UNESCO in Paris. He has attended numerous congresses and conferences as an official delegate, co-organizer, co-chairman, or contributor.

He is a member of many scientific Committees and Organizations, e.g., the Federation of Acoustical Society of Europe (FASE) (Vice-President), the International Council of Science Policy Studies and the Editorial Committe of *Science of Science*. He is also the author of six monographs and about 150 scientific papers.

MAZUR, Marian, b. 1909, graduated from Warsaw Technical University in 1934, became a Doctor of Technical Sciences in 1951, and since 1954 has been Professor in the Electrotechnical Institute.

He started his scientific research work in the domain of electrical heating, moving subsequently to automation and cybernetics. He is now with the Institute of Science Policy in the Ministry of Science, Higher Education and Technology, and his main subject is the application of the cybernetic approach to studies in science policy and research methodology. In 1949–1963 he lectured at Warsaw Technical University and in 1974–1975 at l'École des Hautes Études en Sciences Sociales (Sorbonne)in Paris.

He has been a member of several scientific organizations, e.g., the Committee of the Science of Science of the Polish Academy of Sciences (since 1969); he was also President of Committee No 27 of the International Electrotechnical Commission (in 1960–1971).

He is the author of several books (in Polish), e.g., *Cybernetyczne teorie układów samodzielnych* (*Cybernetic Theory of Autonomous Systems*), 1966; *Jakościowa teoria informacji* (*Qualitative Theory of Information*), 1970 (translated into Russian); *Historia naturalna polskiego naukowca* (*Natural History of Polish Scientists*), 1970; *Cybernetyka i charakter* (*Cybernetics and Character*), 1976.

OLSZEWSKI, Eugeniusz, b. 1911, graduated from Warsaw Technical University in 1935, became a Doctor of Technical Sciences in 1945, and since 1951 has been Professor of this University and head of the Research Group for the Science of Science and Science of Technology.

His main scientific interests are: the history of science and technology, the conceptual apparatus of the science of science, and the science of technology, including the general concept of the scientific and technological revolution.

He has attended many international conferences and congresses, presenting numerous papers.

He is a member or praesidium member of several Committees, Commissions and Societies, e.g., the Polish Academy of Sciences, Committee of the Science of Science, Committee for Research and Prognosis 'Poland 2000', Committee of the History of Science and Technology. He is also a member of the International Academy for the History of Sciences, the International Cooperation in the History of Technology Committee, the Society for the History of Technology Committee, the Society for the History of Technology (USA) and the International Council for Science Policy Studies.

He is editor or a member of the Editorial Board of several scientific journals, e.g., *Problemy Naukoznawstwa*, *Fundamenta Scientiae*.

OSSOWSKA, Maria (1896–1974), sociologist and philosopher, in 1948 became Professor at Warsaw University. She greatly contributed to the development of moral philosophy, particularly in three fields: methods of analysing the character and the foundations of moral evaluations as well as norms, the psychology of morality, and the sociology of morality.

She published several books in Polish and numerous papers both in Polish and English. Her most important books are: *Podstawy nauki o moralności* (*The Foundations of Moral Science*), 1947, IInd ed. 1957; *Motywy postępowania* (*Motives of Conduct*), 1949, IInd ed. 1958; *Oceny i normy* (*Evaluations and Standards*), 1949; *Moralność mieszczańska* (*Bourgeois Morality*), 1956; *Zagadnienie powszechnie uznanych norm moralnych* (*The Problem of Generally Accepted Moral Standards*), 1957; *Socjologia moralności* (*Sociology of Morals*), 1963; *Normy moralne* (*Moral Norms*), 1970; English lang. edition, 1980.

OSSOWSKI, Stanisław (1897–1963), sociologist, theorist of culture, methodologist of the social sciences, in 1947 became Professor at Warsaw University, and in 1956

President of the Polish Sociological Society. From 1959 to 1962 he was Vice-President of the International Sociological Association.

His main interests were: aesthetic value forming one of the aspects of the general theory of culture; criticism of the class-anthropological trend in sociology and the development of humanistic perspectives in sociology; methodological peculiarities of the social sciences.

He published books in Polish and numerous papers either in Polish or in other languages. Published books in Polish: *Podstawy estetyki* (*The Foundations of Aesthetics*), 1933, 1949, 1958, English lang. edition 1980; *Więź społeczna i dziedzictwo krwi* (*Socials Links and Blood Heritage*), 1939, 1948; *O osobliwościach nauk społecznych* (*About the Peculiarities of Social Science*), 1962; *Ku nowym formom życia społecznego* (*Towards New Forms of Social Life*), 1947, 1958.

SZANIAWSKI, Klemens, b. 1925, graduated in mathematics and philosophy at the Universities of Warsaw and Łódź, became Doctor of Philosophy in 1950. He is now Professor of Logic at the Institute of Philosophy, Warsaw University, President of the Polish Philosophical Society (elected in 1977), President of the Polish Committee of the International Union of History and Philosophy of Science (elected in 1973).

Szaniawski is the author of several publications on the logic of non-deductive inference, the theory of decision-making and the philosophy of science.

He is also a member of the editorial boards of *Synthèse*, *Theory and Decision*, *Erkenntnis*, *Philosophy of Science*, *Dialectics and Humanism*, *Poznań Studies in the Philosophy of Sciences and the Humanities*, *Studia Logica*, and *Problemy Naukoznawstwa*.

WALENTYNOWICZ, Bohdan, b. 1912, graduated from Warsaw Technical University in 1936, became a Doctor of Technical Sciences in 1964, and since 1978 has been Professor at the Polish Academy of Sciences, and head of the Science of Science Research Unit (in the Department of Praxiology and the Science of Science in the Institute of Philosophy and Sociology).

His main research interests are: the philosophical, methodological and sociological status of technology and engineering, communication between scientists and engineers, scientific terminology, as well as scientific writing.

For many years he has been editor-in-chief of several scientific journals, e.g. *Zagadnienie Naukoznawstwa*, *Science of Science*, *Przegląd Elektrotechniczny*.

He is the author of several contributions and papers, some of them presented at international conferences and congresses. He is also a member, or Praesidium member, of several scientific organizations, e.g., the Committee of the Science of Science and the Committee of Terminology of the Polish Academy of Sciences, as well as the International Council for Science Policy Studies.

ZNANIECKI, Florian Wiktor (1882–1958), philosopher and sociologist, one of the main representatives of humanistic sociology, studied in Poland, Switzerland and France. In 1920–1939 he was Professor at Poznań University. On his initiative the Polish Sociological Institute was founded and the periodical *Przegląd Socjologiczny* started. At the outbreak of World War II he was in the United States, a country to which he had previously often been invited, and where — initially together with W. I. Thomas — he had carried out research on Polish emigration. In the United States he lectured at the University of Illinois (Urbana) in 1940–1950. In 1953–1954 he was President of the American Sociological Society.

Znaniecki's main scientific interest was sociology. He investigated its intellectual status and scope, and the approach to it and made a great contribution to the transformation of sociology into an academic discipline. He postulated the application of the humanistic approach and the use of empirical data based on documents relating to personal experience. He was also interested in some more detailed problems, such as the sociology of character formation and the sociology of science.

Znaniecki wrote both in Polish and English. The following books were published: *Zagadnienie wartości w filozofii* (*Value Problems in Philosophy*), 1910; *Cultural reality* (1919); *Upadek cywilizacji zachodniej* (*Decadence of Western Civilization*), 1921; *Ludzie teraźniejsi a cywilizacja przyszłości* (*Modern People and Future Civilization*), 1935; *Wstęp do socjologii* (*Introduction to Sociology*), 1922; *The Method of Sociology*, 1934; *Social Actions*, 1936; *The Social Role of the Man of Knowledge*, 1940; *Cultural Sciences*, 1952, and *Social Relations and Social Rôles*, 1965.

INDEX OF NAMES

INDEX OF TERMS

THEORY AND DECISION LIBRARY

An International Series in the Philosophy and Methodology
of the Social and Behavioral Sciences

1. Günther Menges (ed.), *Information, Inference, and Decision.* 1974, viii + 195 pp.
2. Anatol Rapoport (ed.), *Game Theory as a Theory of Conflict Resolution.* 1974, v + 283 pp.
3. Mario Bunge (ed.), *The Methodological Unity of Science.* 1973, viii + 264 pp.
4. Colin Cherry (ed.), *Pragmatic Aspects of Human Communication.* 1974, ix + 178 pp.
5. Friedrich Rapp (ed.), *Contributions to a Philosophy of Technology. Studies in the Structure of Thinking in the Technological Sciences.* 1974, xv + 228 pp.
6. Werner Leinfellner and Eckehart Köhler (eds.), *Developments in the Methodology of Social Science.* 1974, x + 430 pp.
7. Jacob Marschak, *Economic Information, Decision and Prediction. Selected Essays.* 1974, three volumes, xviii + 389 pp.; xii + 362 pp.; x + 399 pp.
8. Carl-Axel S. Staël von Holstein (ed.), *The Concept of Probability in Psychological Experiments.* 1974, xi + 153 pp.
9. Heinz J. Skala, *Non-Archimedean Utility Theory.* 1975, xii + 138 pp.
10. Karin D. Knorr, Hermann Strasser, and Hans Georg Zilian (eds.), *Determinants and Controls of Scientific Developments.* 1975, ix + 460 pp.
11. Dirk Wendt, and Charles Vlek (eds.), *Utility, Probability, and Human Decision Making. Selected Proceedings of and Interdiscyplinary Research Conference, Rome, 3–6 September, 1973,* 1975, viii + 418 pp.
12. John C. Harsanyi, *Essays on Ethics, Social Behavior, and Scientific Explanation.* 1976, xvi + 262 pp.
13. Gerhard Schwödiauer (ed.), *Equilibrium and Disequilibrium in Economic Theory. Proceedings of a Conference Organized by the Institute for Advanced Studies, Vienna, Austria, July 3–5, 1974.* 1978, i + 736 pp.
14. V. V. Kolbin, *Stochastic Programming.* 1977, xii + 195 pp.
15. R. Mattessich, *Instrumental Reasoning and Systems Methodology.* 1978, xxii + + 396 pp.
16. H. Jungermann and G. de Zeeuw (eds.), *Decision Making and Change in Human Affairs.* 1977, xv + 526 pp.

18. A. Rapoport, W. E. Stein, and G. J. Burkheimer, *Response Models for Detection of Change*. 1978 (forthcoming).
19. H. J. Johnson, J. J. Leach, and R. G. Mühlmann (eds.), *Revolutions, Systems, and Theories; Essays in Political Philosophy*. 1978.
20. Stephen Gale and Gunnar Olsson (eds.), *Philosophy in Geography*. 1979, xxii + + 470 pp.
21. Maurice Allais and Ole Hagen (esd.), *Experted Utility Hypotheses and the Allais Paradox: Contemporary Discussions of Decisions Under Uncertainty With Allais' Rejoinder*. 1979, vii + 681 pp. + indexes.
22. Teddy Seidenfeld, *Philosophical Problems of Statistical Inference. Learning from R. A. Fisher*. 1979, xiv + 246 pp.
23. L. Lewin and E. Vedung (eds.), *Politics as Rational Action*. 1980, xii + 274 pp.
24. J. Kozielecki, *Psychological Decision Theory*, 1982, xvi + 395 pp. + index.
25. I. I. Mitroff and R. O. Mason, *Creating a Dialectical Social Science, Concepts, Methods and Models*, 1981, ix + 189 pp.
26. V. A. Lefebvre, *Algebra of Conscience, A Comparative Analysis of Western and Soviet Ethical Systems*, 1982, xxvii + 187 pp. + index (forthcoming).
27. L. Nowak, *Property and Power, Towards a Non-Marxian Historical Materialism*, 1982 (forthcoming).
28. J. C. Harsanyi, *Papers in Game Theory, Papers 1956–1980*, 1982, x + 253 pp. + index (forthcoming).
29. B. Walentynowicz (ed.), *Polish Contributions to the Science of Science*, xii + 280 pp. + indexes.